A.R.T. Autologous Regenerative Therapy in Aesthetic Medicine

Domenico Piccolo • Alessandro Gennai
Fabrizio Melfa • Bruno Bovani • Piero Tesauro

A.R.T. Autologous Regenerative Therapy in Aesthetic Medicine

Standardized Protocols in Tissue Rejuvenation

Springer

Domenico Piccolo
Avezzano, L'Aquila and Pescara
Novea Skin Center
Avezzano, L'Aquila, Italy

Fabrizio Melfa
Aesthetic Medicine
Mediaging Clinic Center
Palermo, Palermo, Italy

Piero Tesauro
Hair Transplant
Private Practice Poliambulatorio Gioia
Milan, Milano, Italy

Alessandro Gennai
Studio Gennai - Bologna- Italy
Milano, MI, Italy

Bruno Bovani
Plastic Surgery
Bovani Surgery And Aesthetic Clinic Center
Perugia, Perugia, Italy

ISBN 978-3-032-05834-8 ISBN 978-3-032-05835-5 (eBook)
https://doi.org/10.1007/978-3-032-05835-5

This Springer imprint is published by the registered company Springer Nature Switzerland AG
The registered company address is: Gewerbestrasse 11, 6330 Cham, Switzerland

I dedicate this work to my wife, Federica, and to my wonderful daughter and son, Camilla and Alex. Their unwavering support and encouragement continually inspire me to strive beyond my limits.

Domenico Piccolo

To my daughter
"Dear Blu,
Only by having the courage to lose the sight of the shore you can discover new Oceans"
With love

Alessandro Gennai

To my daughters, Sonia and Giulia
"May passion and curiosity always be your guiding stars, in life and in the scientific careers you have chosen."
With love and admiration.

Fabrizio Melfa

I wish to dedicate the chapter I have written to two people: my father Franco and my son Saverio.
My father is no longer with us, but I am certain he would have been deeply proud of me and of the path I have followed thus far. I

must thank him for always giving me the right guidance so that I would never forget the main objective to pursue: to care for people and improve their well-being.
I also dedicate this chapter to my son Saverio, who has chosen to follow in my footsteps and is proving to be a skilled surgeon.
Regenerative Autologous Medicine has been a remarkable discovery for our generation, and day by day it continues to open up new horizons. But for my son's generation, I am sure its use will have become second nature. And I am equally sure that, in time, my son too will remember the small scientific contribution I made to the development of this extraordinary field.
I also thank Dr. Monica Baldessin for her valuable support to the chapter dedicated to the synergy of A.R.T. with EBD. I am sure that our vision of a combined action between the use of our mesenchymal cells and some treatments aimed at tissue regeneration will increasingly represent an essential reality for our profession.

Bruno Bovani

To Gabry, for everything I took away from her.

Piero Tesauro

Preface

Regenerative medicine is a field that has seen remarkable progress over the last two decades, continually providing new tools for physicians to restore and improve patients' quality of life. Within this dynamic landscape, adipose tissue has emerged as a central resource, not only for its abundance but also for its regenerative potential.

It was during our ongoing collaboration and exchange of clinical experiences that the idea of this book began to take shape. We shared a common vision: to refine adipose tissue harvesting techniques in a way that would make regenerative procedures safer, more standardized, and accessible to a broader range of practitioners. Out of this vision came the Guided SEFFI (Superficial Enhanced Fluid Fat Injection) technique, an evolution of the SEFFI method that some of us had previously standardized and published. Guided SEFFI introduces a decisive innovation by allowing tissue to be collected at precise dimensions during the harvesting phase, avoiding unnecessary manipulation and thus preserving cellular vitality.

This advancement not only ensures higher consistency in results but also simplifies the procedure, enabling physicians without extensive surgical training to perform it safely and effectively. We believe that this democratization of access to regenerative therapy is one of the most important contributions of Guided SEFFI, and it is our hope that this book will help to disseminate its principles and practice to a wider audience.

The applications of Autologous Regenerative Therapy are already proving valuable in aesthetic medicine and trichology, which are the main focus of this manual. Yet the potential reaches much further: promising results are emerging in orthopedics, gynecology, wound care, urology, and other fields where tissue repair and regeneration may profoundly change clinical practice.

Looking forward, regenerative medicine will continue to evolve through clinical trials, collaborative research, and the refinement of protocols. We trust that Guided

SEFFI will contribute to this evolution by offering a reliable, minimally invasive, and reproducible approach. Ultimately, the true measure of progress will be found in improved patient outcomes and enhanced quality of life.

Avezzano, L'Aquila, Italy	Domenico Piccolo
Bologna, Milano, Italy	Alessandro Gennai
Palermo, Palermo, Italy	Fabrizio Melfa
Perugia, Perugia, Italy	Bruno Bovani
Milan, Milano, Italy	Piero Tesauro

Note on AI Usage

All content is original and authored by the contributors, and has not been generated by Artificial Intelligence. However, AI was used to check grammatical accuracy and to generate abstracts and keywords, with the aim of appropriately reflecting the authors' original content within the chapters.

Acknowledgements

Special thanks to Matteo Migliori who coordinated the writing of this book from day one. Thanks to his scientific and relational skills: he immediately understood the importance of this manual for the doctor who wants to bring Autologous Regenerative Therapy into his practice and offer his patients safe, effective, and standardized tissue regeneration.

Contents

Chapter 1
The A.R.T.: Autologous Regenerative Therapy Approach for Rejuvenating Tissues

1.1 Introduction: Tissue and Skin Senescence

The skin, being the largest organ of the human body, serves a dual role as a protective shield against external environmental factors and a significant contributor to our overall appearance. As we age, the skin undergoes a complex and multifaceted process known as skin aging, characterized by a range of physiological and structural changes. A pivotal factor in skin aging is the progressive reduction in the extracellular matrix (ECM), a dynamic network of macromolecules crucial for maintaining skin health and functionality [1, 2].

Skin aging is a phenomenon influenced by both intrinsic and extrinsic factors. Intrinsic aging, often referred to as chronological aging, is an inevitable and genetically determined process that occurs over time. Extrinsic aging, conversely, results from external factors such as UV radiation, pollution, lifestyle choices, and other environmental stressors. Both intrinsic and extrinsic aging processes converge on the ECM, leading to its gradual degradation and alteration.

The ECM comprises major components, including collagen, elastin, fibronectin, and hyaluronic acid, which collectively play a significant role in determining skin texture, firmness, and overall appearance. Central to the aging process of the skin are metalloproteinases (MMPs), a family of enzymes responsible for maintaining the balance of the ECM. MMPs are crucial for the degradation and remodeling of various ECM components, including collagen, elastin, and fibronectin [1, 3].

In youthful, healthy skin, MMPs are tightly regulated to uphold the structural integrity of the ECM, ensuring proper tissue function and a youthful appearance. However,

Supplementary Information The online version contains supplementary material available at https://doi.org/10.1007/978-3-032-05835-5_1.

D. Piccolo et al., *A.R.T. Autologous Regenerative Therapy in Aesthetic Medicine*, https://doi.org/10.1007/978-3-032-05835-5_1

during the aging process, the equilibrium between MMPs and their tissue inhibitors (TIMPs) becomes disrupted, leading to excessive and uncontrolled MMP activity. This deregulation of MMPs contributes significantly to the degradation of ECM constituents, particularly collagen, which serves as a critical structural protein in the skin [1, 4, 5].

Furthermore, the aging process is exacerbated by the presence of senescent fibroblast populations within the skin. These senescent cells do not undergo apoptosis or removal by the adaptive immune system but instead persist within the tissue and release a secretory phenotype known as SASP (senescence-associated secretory phenotype). This SASP creates a pro-aging microenvironment, activating MMPs and leading to further breakdown of the ECM while suppressing the release of growth factors necessary for tissue regeneration. The marked decrease in collagen over time may result from factors beyond the diminishing populations of fibroblasts, with negative feedback loops being created as a consequence of high molecular weight collagen fragments.

In summary, skin aging is a multifaceted process influenced by intrinsic and extrinsic factors, with the progressive degradation of the ECM and the dysregulation of MMPs playing central roles. The presence of senescent fibroblasts and the release of SASP further exacerbate these effects, contributing to the loss of skin firmness, texture, and overall youthful appearance. Understanding the intricate mechanisms of skin and tissue senescence is crucial for developing targeted regenerative therapies that aim to rejuvenate aging skin and restore its vitality.

1.2 Autologous Regenerative Therapy (ART) in Medicine and Aesthetics

Autologous homologous regenerative therapy has emerged as a transformative frontier in regenerative medicine, offering a promising approach to tissue repair and regeneration. This innovative therapeutic approach capitalizes on the regenerative potential of a patient's own cells and tissues, ensuring compatibility and reducing the risk of immune rejection and adverse reactions associated with foreign substances. By meticulously isolating and processing these autologous cells, they can be reintroduced into the target area to stimulate tissue regeneration, repair damaged structures, and enhance the natural healing processes of the body.

The potential applications of autologous homologous regenerative therapy span a wide range of medical conditions, making it a versatile and promising field within regenerative medicine.

However, in this book, we focus our attention on a specific aspect of this groundbreaking therapy—autologous homologous regenerative therapy utilizing adipose tissue.

1.3 Adipose Tissues: Their Role and Classification

Adipose tissues, abundant in adult stem cells, have gained considerable attention in scientific research due to their role in various physiological functions. Derived from the mesenchyme, adipose tissues are intricate and comprise a myriad of cells: stem cells, endothelial cells, collagen, resident monocytes/macrophages, lymphocytes, fibroblasts, vascular smooth muscle cells, preadipocytes, and adipocytes. These tissues are not just storage depots but are active players in homeostasis, metabolism regulation, and even aging processes.

In mammals, adipose tissues can be broadly categorized into two main types: white adipose tissue (WAT) and brown adipose tissue (BAT). Both WAT and BAT are equipped with lipolytic and lipogenic functions, pivotal for energy storage and release [6].

BAT, enriched with vascularization, cytochromes, and mitochondria, specializes in energy dissipation. In fetuses and neonates, BAT is predominantly located in the axillary, perirenal, periadrenal, and cervical regions. Intriguingly, as mammals mature, most of this BAT transforms into WAT [7, 8].

WAT, on the other hand, can be found in both subcutaneous and visceral depots. Unique in its characteristics, WAT, especially in its youthful state with a high turnover of adipocytes, is not directly linked to metabolic disorders. This tissue is dispersed across multiple sites in humans, including the subcutaneous regions of the abdomen, thighs, and buttocks, as well as in the intestines, perirenal areas, omentum, retro-orbital spaces, and bone marrow [9–11].

1.4 White Adipose Tissue (WAT) in Regenerative Therapy

White adipose tissue (WAT) is a particularly attractive candidate for autologous homologous regenerative therapy for several compelling reasons [12].

Firstly, it is abundantly available in the human body and easily accessible, eliminating the need for invasive procedures to obtain donor tissue. This not only reduces patient discomfort but also minimizes the potential for complications associated with donor tissue procurement. Secondly, adipose tissue is inherently homologous, originating from the same individual, thereby significantly reducing the risk of immune rejection and graft-versus-host reactions.

Central to the success of regenerative therapy using adipose tissue are the cells of the stromal vascular fraction (SVF) and their secretomes and exosomes. SVF is a complex and heterogeneous population of cells derived from adipose tissue, including adipose-derived stem cells (ADSCs), endothelial cells, pericytes, and immune cells, among others. These cells collectively contribute to the regenerative potential of adipose tissue-based therapies. ADSCs, in particular, are highly versatile and can differentiate into various cell types crucial for tissue repair, such as adipocytes, osteoblasts, and chondrocytes.

Additionally, the secretomes and exosomes released by these cells contain a rich cocktail of growth factors, cytokines, and microRNAs. These bioactive molecules serve as potent signaling molecules, influencing neighboring cells to promote tissue regeneration, reduce inflammation, and enhance healing processes. Harnessing the regenerative capabilities of SVF and their secreted factors represents a powerful approach in regenerative medicine, with far-reaching potential applications.

1.5 Distinct Facets: ARTM and ARTA

In the field of regenerative medicine, Autologous Regenerative Therapy in Medicine (ARTM) and Autologous Regenerative Therapy in Aesthetic (ARTA) represent two distinct yet interconnected facets, each with its own set of goals and considerations [13–16].

ARTM primarily focuses on the restoration of functionality in tissues, organs, or body parts that have been compromised or damaged due to trauma, disease, or degenerative conditions. This field encompasses a wide range of applications, including strategies for wound healing, tissue repair, and addressing disease-state tissue environments. The core objective of ARTM is to improve the overall health and well-being of patients by promoting the recovery and regeneration of vital bodily structures. To achieve this, researchers and practitioners in ARTM draw upon a diverse toolkit of regenerative approaches tailored to specific medical conditions, emphasizing the need for precision and individualized treatment strategies.

Conversely, the primary goal of Autologous Regenerative Therapy in Aesthetic (ARTA) is to address the aesthetic concerns of individuals by focusing on the regeneration of soft tissue that may have been lost or damaged due to the natural aging process or other factors. Unlike many ARTM strategies that deal with wound healing or disease-state tissue environments, the predominant environment in ARTA is senescent tissue. This distinction is crucial, as it underscores the unique challenges and opportunities in the field of aesthetic regenerative therapy. Practitioners in ARTA need to be particularly attuned to the specific characteristics of senescent tissue, tailoring their regenerative approaches accordingly to optimize outcomes in the pursuit of aesthetic rejuvenation.

1.6 Scope and Focus of this Book

It is important to note that, while the scope of Autologous Regenerative Therapy (ART) is vast and extends across numerous medical fields, this book focuses exclusively on its applications in aesthetic medicine. This specialization is driven by the recognition that regenerative medicine is a multifaceted and intricate field, with each medical specialty demanding its own unique considerations, protocols, and evidence-based practices. The decision to concentrate on aesthetics is not a limitation but rather a deliberate choice to provide in-depth exploration of a specific facet of ART, offering colleagues in the aesthetics domain valuable insights and knowledge.

However, it is essential for readers to acknowledge the broader scope of ART's potential across various medical fields, including but not limited to orthopedics, cardiology, neurology, wound care, proctology, gynecology, dermatology, and more. For those seeking information and guidance on specific non-aesthetic applications of Autologous Regenerative Therapy, we recommend exploring dedicated resources in those respective areas. In this book, we aim to provide a comprehensive understanding of the intricacies and advancements in Autologous Regenerative Therapy within the field of aesthetics, empowering practitioners with the knowledge they need to offer cutting-edge treatments and enhance patient outcomes.

1.7 Adipose Tissue Transplantation and Modern Techniques

1.7.1 The Genesis of Adipose Tissue Studies

At the intersection of the nineteenth and twentieth centuries, the pioneering studies on adipose tissue transplantation began. The focus was predominantly on reconstructive applications. Lexer in 1910 and Hollander in 1912 highlighted the use of adipose tissue for correcting Romberg's syndrome and repairing neoplasm excision defects. However, these early efforts faced challenges like a poor engraftment rate, which often hovered between 20–50%. The low engraftment rate necessitated overcorrection to approach the desired outcome.

1.7.2 The Evolution of Techniques

During the subsequent years, the medical community, led by influential figures like Gurney in 1938 and Peer in 1959, pivoted towards minimizing trauma during transplantation. They observed that atraumatic manipulation correlated positively with improved graft uptake. However, tissue extraction methods remained rudimentary, often using dermo-adipose samples that were later refined. Due to limitations in extraction techniques, excessively large samples were often harvested, posing challenges for graft survival.

1.7.3 The Emergence of Liposculpture

The advent of liposculpture marked a significant shift in adipose transplantation. The introduction of cannulas and syringes facilitated the less traumatic extraction of vast adipose tissue amounts. Illouz and Fournier, by the mid-1980s, began repurposing adipose tissue extracted during liposculpture to rectify irregularities stemming from the procedure or previous surgeries. Other proponents of this method, such as Gasparotti, Lewis, and Toledo, further refined the technique.

1.7.4 The Lipostructure Revolution

1997 heralded the advent of Lipostructure, also coined the “Coleman technique.” Coleman pioneered and standardized this method, focusing on efficient fat grafting, particularly for facial treatments [17].

The essence of Coleman’s technique rested on three pillars:

Gentle harvesting with blunt cannulas.
Efficient centrifugation.
Regrafting using finer cannulas.

The commitment to minimizing trauma remained consistent, with emphasis on employing 10 cc syringes for harvesting at low pressures, centrifuging the adipose tissue at 3000 g for 3 min, and regrafting using a multi-planar and multi-tunnel technique.

1.7.5 Refining the Coleman Technique

The technique introduced by Coleman has since witnessed myriad modifications, from changes in cannula sizes to centrifugation debates. The outcomes of lipostructure were not just confined to volumizing effects. Several practitioners, including Coleman, reported prolonged skin quality and structure improvements post-procedure. In 2007, Rigotti’s work in treating radiodermic tissues propelled further interest in understanding mesenchymal stem cells’ properties [18].

1.7.6 The Paradigm Shift to Stem Cells

Increasingly, the research narrative pivoted from adipocytes to stem cells. Focus sharpened on the differentiation and regenerative capacities of ADSCs, leading to refinements in extraction techniques, centrifugation processes, and grafting methodologies.

1.7.7 Contemporary Techniques: Understanding SEFFI

With the classic lipofilling technique, it was observed that the grafted volume diminished rather quickly and unpredictably, while skin quality and tone continued to improve over months. Additionally, adipose tissue harvested using cannulas with large apertures (often >2 mm) consisted of large lobules. This made certain areas, like the periocular zone, almost untreatable due to the risk of contour irregularities. In some regions, deep grafting was impractical because of extensive dissection if paired with surgical techniques.

The natural evolution of lipofilling led to the micro and nanofat graft techniques [19–21].

By "fluidifying" the tissue, it enabled a more superficial implantation with advantages as follows:

- Enhanced adipose tissue graft survival.
- Achieving a more fluid and uniform tissue for superficial grafting.
- Amplifying the regenerative potential of adipose tissue.

For the first advantage, it is understood that the more distributed the implanted tissue is and the more superficial the graft, the greater the regenerative effect on the skin and toning of the subcutaneous layer. To achieve this, a fluid tissue composed of small cellular "clusters" that do not cause surface irregularities is essential. For the third advantage, it is recognized that skin quality improvements are due to the action of the SVF (stromal vascular fraction) present in the aspirated adipose tissue, as previously discussed.

The stromal vascular fraction (SVF) and adipose derived stem cells (ADSCs), with their paracrine actions, are the valuable "gold" found within the "river" of harvested superficial subcutaneous tissue.

Using this tissue graft merely as a "filler" would be analogous to utilizing "stones" instead of "gold." Our research focused on perfecting a technique to maximize this tissue's potential: restoring volume and tissue regeneration, with the latter being uniquely paramount.

With this premise, it is essential to understand how the Superficial Enhanced Fluid Fat Injection (SEFFI) technique differs from the micro and nanofat graft techniques [22–25].

The challenge with micro and nanofat graft, which we aimed to overcome with the SEFFI technique, is as follows:

The skin and subcutaneous layer of the periocular region and face vary in thickness and characteristics across areas. It is essential to have tissues with varying fluidity and cellularity to use superficially in ALL facial and periocular regions. Harvesting with large-aperture cannulas (>1 mm in diameter) yields larger cellular clusters that require significant manipulation using devices aiming solely to mechanically reduce cluster sizes. This mechanical manipulation inevitably causes cellular damage, reduces cellular viability, results in tissue loss trapped in the device, prolongs the procedure, and, as demonstrated in nanofat graft, leads to the complete destruction of mature adipocytes [21].

1.8 Standardization and Best Practices in Regenerative Autologous Therapy: SEFFI and Guided SEFFI

As previously mentioned, given the widespread adoption of Autologous Regenerative Therapy (ART) across a vast array of medical fields (aesthetic medicine, plastic surgery, trichology, orthopedics, gynecology, wound care, proctology), it becomes imperative to ensure a strict standardization of the procedure for extracting and preparing adipose tissue. This allows physicians from all specialties to perform it in an outpatient setting, simply, safely, and effectively. Based on recent scientific evidence, the standardized procedure must have the following characteristics:

- Enable tissue extraction in the least traumatic way possible, using syringe extraction and low vacuum pressure [26–30].
- Harvest adipose tissue from the most superficial plane to obtain tissue richer in stromal vascular fraction (SVF) cells: superficial adipose tissue (SAT) has a higher concentration of SVF cells [31–33, 35].
- Wash and concentrate tissue by decanting in order to obtain the highest ADSC concentration and washing out the lidocaine [34–37].
- Extract cellular clusters containing adipocytes, stromal tissue (SVF), and mesenchymal stem cells (ADSCs) that are sufficiently small, allowing the tissue to be injected with a needle or microcannula into the superficial subcutaneous tissue, facilitating engraftment in the recipient site [38–40].
- Limit the manipulation of the harvested tissue to maintain the highest possible cellular vitality and stemness, avoiding the use of tissue fragmentation devices [41, 42].
- Minimize trauma for the patient and reduce extraction-related complications such as hematomas, seromas, infections, and skin irregularities [43].
- Following these key points, we standardized and published guided Superficial Enhanced Fluid Fat Injection method for Autologous Regenerative Therapy in Aesthetics (ARTA) [44].

This aims to make the technique simple, safe, effective, and standardized for aesthetic doctors, dermatologists, trichologists, and plastic surgeons. The method was assessed in a multicenter Italian study involving 2365 cases and demonstrated a high degree of efficacy and safety, even when combined with other aesthetic medicine techniques [45].

In the guided SEFFI technique, the use of a guide is foreseen to standardize both the administration of the anesthetic solution and the extraction of adipose tissue clusters at a maximum depth of 15 mm. This makes the technique simple, safe, and effective even for doctors without specific expertise in extracting adipose tissue.

The guided microcannula system (Fig. 1.1), a pivotal component of the SEFFI technique, is ingeniously designed to enhance both safety and precision during the procedure. This system is characterized by a specific spatial arrangement where the microcannula is positioned 15 mm away from the guide's blade, and its tip protrudes exactly 15 mm beyond this point. Such a configuration ensures that during the critical steps of anesthetic administration and adipose tissue harvesting, the cannula's penetration depth beneath the skin is consistently limited to no more than 15 mm. Furthermore, this design facilitates the collection of adipose tissue exclusively from the superficial layer, not exceeding a depth of 15 mm.

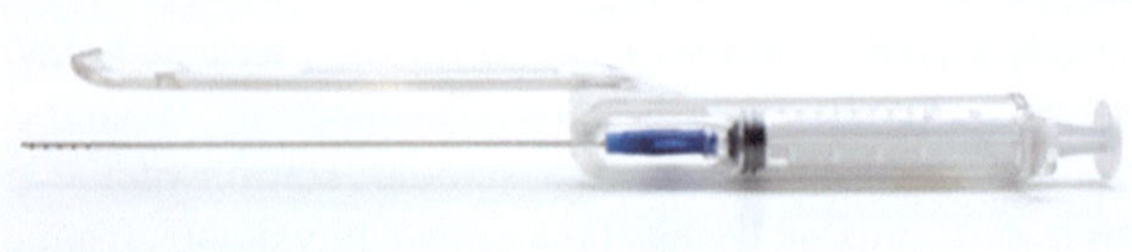

Fig. 1.1 The syringe and cannula inserted in the guide

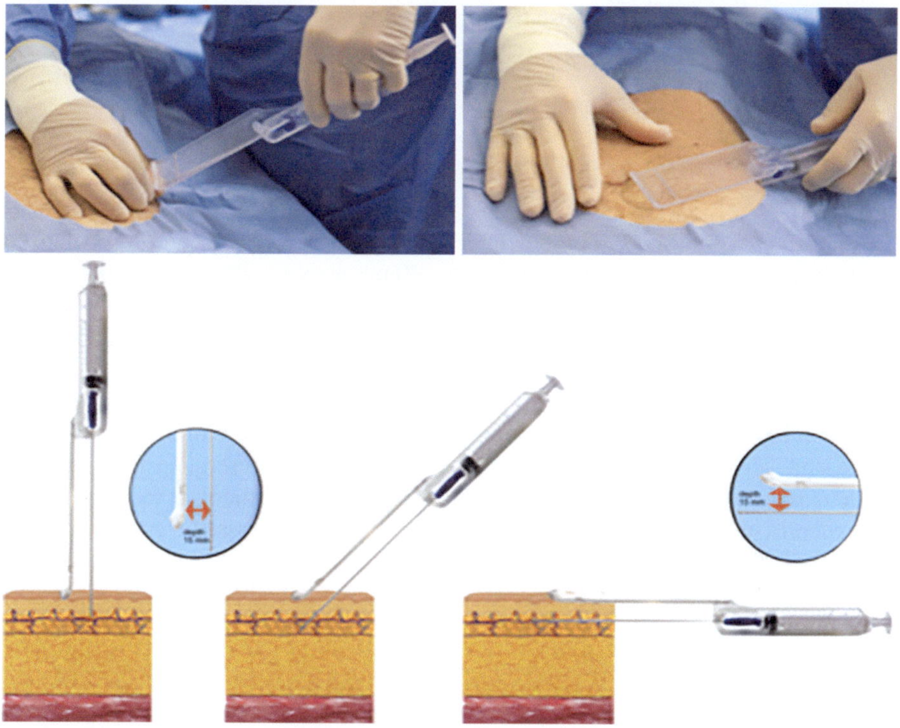

Fig. 1.2 The guided SEFFI technique: the guide prevents the cannula from penetrating more than 15 mm, and direct it to harvest at 15mm depth, ensuring standardized harvesting of the Superficial Adipose Tissue

One of the most significant advantages of this guided system is its accessibility to physicians who may not possess advanced surgical skills. By employing this guided approach, even less experienced practitioners can safely and effectively administer the anesthetic solution and harvest superficial adipose tissue (SAT). This feature greatly broadens the scope of professionals who can competently perform the SEFFI procedure, ensuring it is conducted in a manner that is safe, simple, and efficient. This innovation in the SEFFI technique (Fig. 1.2) thus represents a significant advancement in the field of facial Autologous Regenerative Therapy.

1.9 Detailed Protocol for Anesthetic Administration and Adipose Tissue Harvesting in Guided SEFFI Procedures

The Guided Superficial Enhanced Fluid Fat Injection (SEFFI) procedure represents a sophisticated approach in facial Autologous Regenerative Therapy, particularly focusing on the precise administration of local anesthesia and the meticulous harvesting of adipose tissue. This procedure is primarily performed under local anesthesia and can utilize adipose tissue from various body areas, with a preference for the abdomen and trochanteric regions due to their abundant fat depots.

1.9.1 Preparation and Administration of Anesthetic Solution

The anesthetic solution used in the Guided SEFFI procedure is a carefully balanced mixture, comprising 10 ml of 2% lidocaine, 30 ml of saline solution, and 0.4 mg/ml of adrenaline. The application of this solution is integral to the procedure and is delivered using the same guide, 10 mL Vaclok and cannula that used for harvesting procedure (Fig. 1.3).

To initiate the process, a minor incision is made in the skin with an 18G needle. The cannula is then inserted perpendicularly into this incision until it reaches the guide stop. This is followed by rotating the cannula by 90 degrees, ensuring that the guide is parallel and adherent to the skin surface. The anesthetic is then infiltrated with a consistent back-and-forth motion of the cannula across the entire intended harvesting area. A waiting period of approximately 10 min is recommended post-anesthesia administration to ensure effective local numbing before commencing the tissue harvesting.

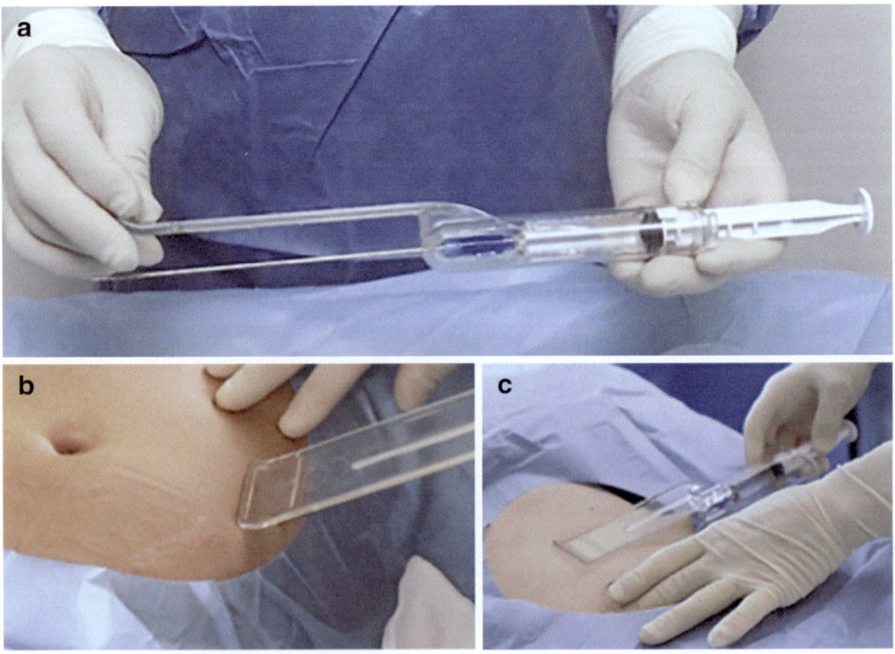

Fig. 1.3 Anesthetic solution administration: (**a**) syringe with anesthetic solution inserted into the guide. (**b**) 15 mm penetration of the tip of the cannula. (**c**) in injection of the anesthetic solution in the superficial adipose tissue (SAT)

1.9.2 Adipose Tissue Harvesting Procedure

The adipose tissue harvesting mirrors the anesthetic administration in terms of the equipment used. The procedure starts by inserting the prepared Vaclok syringe with the cannula into the guide, followed by the insertion of the cannula tip perpendicularly through the skin incision until the guide stops. The cannula is then rotated by 90 degrees to align the guide parallel to the skin, positioning the cannula correctly in the subcutaneous plane. The syringe plunger is pulled back and rotated clockwise to lock and create a vacuum within the syringe barrel.

Maintaining the guide parallel to the skin, the cannula is moved back and forth in a fanning motion across the entire harvesting area. This motion continues until about 5 ml of tissue is collected in the syringe. The cannula is then withdrawn from the skin, and the syringe is detached. The harvested tissue is transferred to a reservoir syringe using a luer lock transfer connector (Fig. 1.4).

1.9.3 Tissue Preparation and Injection

Once in the reservoir syringe, the tissue is mixed with sterile saline solution and left to separate by gravity in a specific syringe holder. This separation process results in the stratification of the tissue (upper layer) from the saline solution (lower layer).

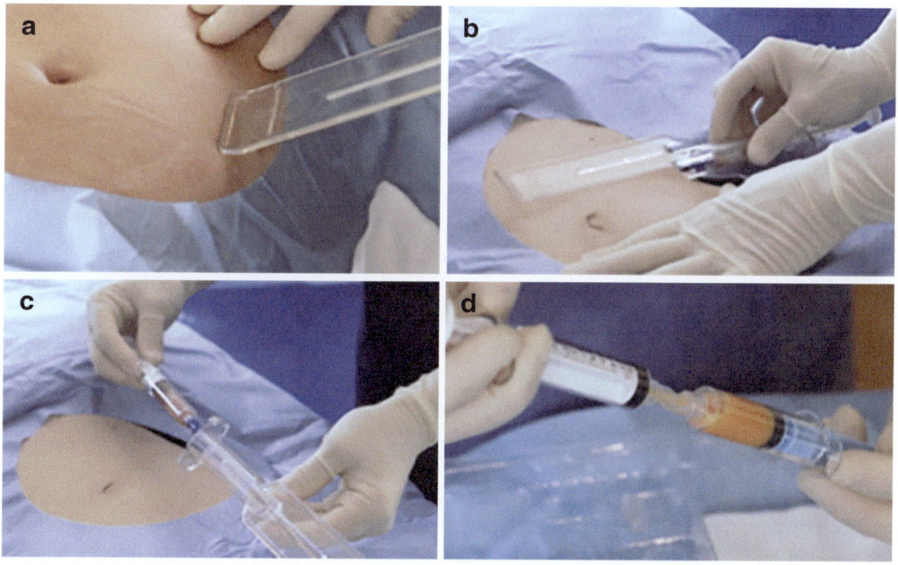

Fig. 1.4 Harvesting procedure of SAT: (**a**) penetration of the tip of the cannula. (**b**) cannula harvesting the SAT. (**c**). harvested adipose tissue into the syringe. (**d**) transferring the tissue from the harvesting Syringe to the decanting one

After a few minutes, the syringe is removed from the holder and the saline solution is discarded, retaining the adipose tissue.

For injection, the prepared tissue is transferred to a 3 ml syringe, excluding the oily upper layer. This refined tissue, now termed microfat, is ideal for injection with a 18G cannula. If the treatment indication requires more fluid tissue, known as nanofat, the tissue is further processed by transferring it between syringes to achieve the desired consistency (Fig. 1.5).

This comprehensive protocol illustrates the nuanced and detailed steps involved in the guided SEFFI procedure, emphasizing the precision required to ensure safety and efficacy. The subsequent chapters will delve into the nuances of tissue processing, differentiating between microfat and nanofat, and their specific applications in facial rejuvenation treatments.

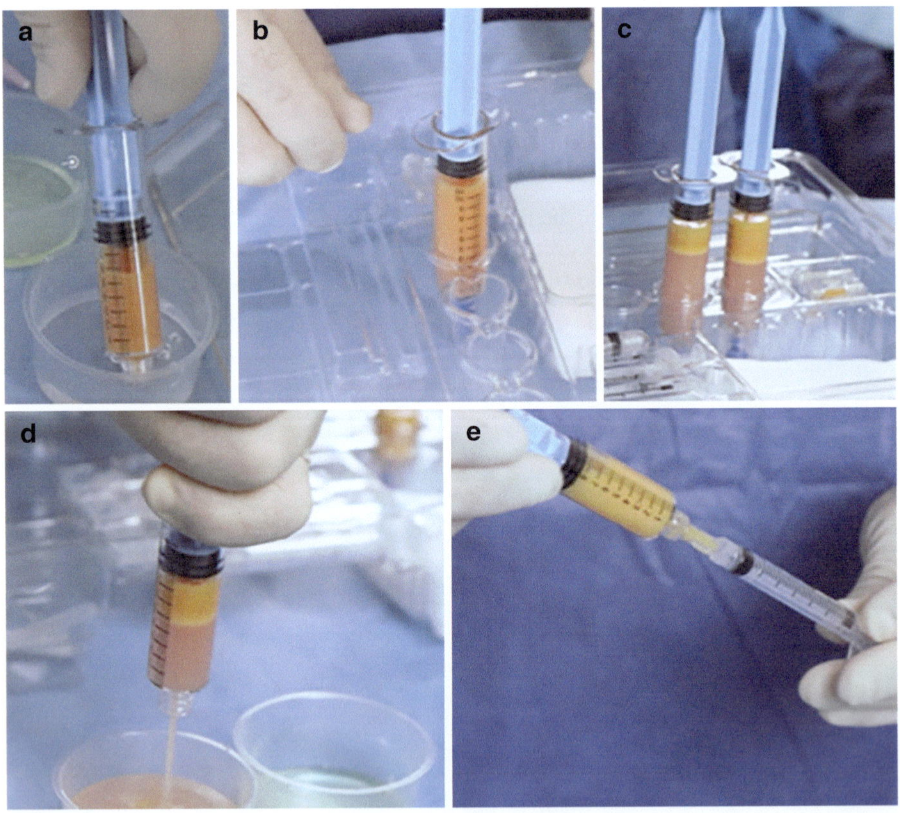

Fig. 1.5 Harvested adipose tissue preparation: (**a**). filling up the syringe with saline solution. (**b**) insertion of the capped syringe into the decanting slot (**c**). after 3–4 min the tissue separates from the washing liquid (**d**) removing the washing liquid. (**e**) transfer the washed tissue into the smaller syringe for injection

References

1. Shin JW, Kwon SH, Choi JY, Na JI, Huh CH, Choi HR, Park KC. Molecular mechanisms of dermal aging and antiaging approaches. Int J Mol Sci. 2019;20(9):2126. https://doi.org/10.3390/ijms20092126. PMID: 31036793
2. Levi N, Papismadov N, Solomonov I, Sagi I, Krizhanovsky. The ECM path of senescence in aging: components and modifiers. V FEBS J. 2020;287(13):2636–46. https://doi.org/10.1111/febs.15282. Epub 2020 Mar 20.PMID: 32145148
3. Cabral-Pacheco GA, Garza-Veloz I, Castruita-De la Rosa C, Ramirez-Acuña JM, Perez-Romero BA, Guerrero-Rodriguez JF, Martinez-Avila N, Martinez-Fierro ML. The roles of matrix Metalloproteinases and their inhibitors in human diseases. Int J Mol Sci. 2020;21(24):9739. https://doi.org/10.3390/ijms21249739. PMID: 33419373.
4. Lee H, Hong Y, Kim M. Structural and functional changes and possible molecular mechanisms in aged skin. Int J Mol Sci. 2021;22(22):12489. https://doi.org/10.3390/ijms222212489. PMID: 34830368.
5. Rousselle P, Montmasson M, Garnier C. Matrix extracellular matrix contribution to skin wound re-epithelialization. Biol. 2019;75–76:12–26. https://doi.org/10.1016/j.matbio.2018.01.002. Epub 2018 Jan 10. PMID: 29330022.
6. Casteilla L, Dani C. Adipose tissue-derived cells: from physiology to regenerative medicine. Diabetes Metab. 2006;32(Pt 1):393–401.
7. Nedergaard J, Bengtsson T, Cannon B. Unexpected evidence for active brown adipose tissue in adult humans. Am J Physiol Endocrinol Metab. 2007;293:E444–52.
8. Gesta S, Tseng Y-H, Kahn CR. Developmental origin of fat: tracking obesity to its source. Cell. 2007;131:242–56.
9. Bjorntorp P, Bengtsson C, Blohme G, Jonsson A, Sjostrom L, Tibblin E, Tibblin G, Wilhelmsen L. Adipose tissue fat cell size and number in relation to metabolism in randomly selected middle-aged men and women. Metabolism. 1971;20:927–35.
10. Salans LB, Cushman SW, Weismann RE. Studies of human adipose tissue adipose cell size and number in nonobese and obese patients. J Clin Invest. 1973;52:929–41.
11. Berry DC, Stenesen D, Zeve D, Graff JM. The developmental origins of adipose tissue. Development. 2013;140:3939–49.
12. Zuk PA, Zhu M, Ashjian P, De Ugarte DA, Huang JI, Mizuno H, et al. Human adipose tissue is a source of multipotent stem cells. Mol Biol Cell. 2002;13:4279–95. https://doi.org/10.1091/mbc.e02-02-0105.
13. Bellei B, Migliano E, Picardo M. Research update of adipose tissue-based therapies in regenerative dermatology. Stem Cell Rev Rep. 2022;18(6):1956–73. https://doi.org/10.1007/s12015-022-10328-w. Epub 2022 Mar 1. PMID: 35230644.
14. Gaur M, Dobke M, Lunyak VV. Mesenchymal stem cells from adipose tissue in clinical applications for dermatological indications and skin aging. Int J Mol Sci. 2017;18(1):208. https://doi.org/10.3390/ijms18010208. PMID: 28117680
15. Wolf DA, Beeson W, Rachel JD, Keller GS, Hanke CW, Waibel J, Leavitt M, Sacopulos M. Mesothelial stem cells and stromal vascular fraction for **skin** rejuvenation. Facial Plast Surg Clin North Am. 2018;26(4):513–32. https://doi.org/10.1016/j.fsc.2018.06.011. Epub 2018 Aug 16. PMID: 30213431.
16. Wang J, Chen Y, He J, Li G, Chen X, Liu H. Anti-**Aging** Effect of the Stromal Vascular Fraction/Adipose-Derived Stem Cells in a Mouse Model of **Skin Aging** Induced by UVB Irradiation. Front Surg. 2022;9:950967. https://doi.org/10.3389/fsurg.2022.950967. eCollection 2022.PMID: 35874134.
17. Coleman SR. Facial recontouring with lipostructure. Clin Plast Surg. 1997;24:347–67.
18. Rigotti G, Marchi A, Galiè M, Baroni G, Benati D, Krampera M, Pasini A, Sbarbati A. Clinical treatment of radiotherapy tissue damage by lipoaspirate transplant: a healing process mediated by adipose-derived adult stem cells. Plast Reconstr Surg. 2007;119(5):1409–22. https://doi.org/10.1097/01.prs.0000256047.47909.71. PMID: 17415234.

19. Trepsat F. Midface reshaping with micro-fat grafting (in French). Ann Chir Plast Esthet. 2009;54:435–43.
20. Zeltzer AA, Tonnard PL, Verpaele AM. Sharp-needle intradermal fat grafting (SNIF). Aesthet Surg J. 2012;32:554–61.
21. Tonnard PL, Verpaele AM, Peeters G et al. Nanofat grafting: basic research and clinical applications. Plast Reconstr Surg J 2013;132:1017–1026.
22. Bernardini FP, Gennai A, Izzo L, et al. Superficial enhanced fluid fat injection (SEFFI) to correct volume defects and skin aging of the face and periocular region. Aesthet Surg J. 2015;35(5):504–15.
23. Gennai A, Zambelli A, MD, Repaci E, Quarto R, Baldelli I, Fraternali G, Bernardini F. P, Skin rejuvenation and volume enhancement with the micro superficial enhanced fluid fat injection (M-SEFFI) for skin aging of the periocular and perioral regions. Aesthet Surg J. 2016:1–10.
24. Gennai FP, Bernardini. Superficial enhanced fuid fat injection (SEFFI and MicroSEFFI) in facial rejuvenation. CellR4. 2017;5(1):e2239.
25. Rossi M, Roda B, Zia S,Vigliotta I, Zannini C, Alviano F, Bonsi L, Zattoni A,Reschiglian P, Gennai A, Characterisation of the tissue and stromal cell components of micro-superficial enhanced fluid fat injection (micro-SEFFI) for facial aging treatment. Aesthet Surg J. 2018:1–12.
26. Cucchiani R, Corrales L. The effects of fat harvesting and preparation, air exposure, obesity, and stem cell enrichment on adipocyte viability prior to graft transplantation. Aesthet Surg J. 2016;36(10):1164–73. https://doi.org/10.1093/asj/sjw106. Epub 2016 Jul 29. PMID: 27474770.
27. Cervelli D, Gasparini G, Moro A, Pelo S, Foresta E, Grussu F, D'Amato G, De Angelis P, Saponaro G. Lipofilling as refinement procedure in maxillo-mandibular malformations. Acta Otorhinolaryngol Ital. 2016;36(5):368–72. https://doi.org/10.14639/0392-100X-857. PMID: 27958596; PMCID: PMC5225791.
28. Vermeersch N, De Fré M, Verhoeven V, Hunter JE, Tondu T, Thiessen FEF. Autologous fat transfer for hand rejuvenation: a systematic review on technique and outcome. JPRAS Open. 2022;32:182–94. https://doi.org/10.1016/j.jpra.2022.03.001. PMID: 35449731; PMCID: PMC9018121.
29. Chou CK, Lee SS, Lin TY, Huang YH, Takahashi H, Lai CS, Lin SD, Lin TM. Micro-autologous fat transplantation (MAFT) for forehead Volumizing and contouring. Aesth Plast Surg. 2017;41(4):845–55. https://doi.org/10.1007/s00266-017-0883-2. Epub 2017 Apr 27. PMID: 28451799; PMCID: PMC5522520.
30. Smith P, Adams WP Jr, Lipschitz AH, Chau B, Sorokin E, Rohrich RJ, Brown SA. Autologous human fat grafting: effect of harvesting and preparation techniques on adipocyte graft survival. Plast Reconstr Surg. 2006;117(6):1836–44. https://doi.org/10.1097/01.prs.0000218825.77014.78. PMID: 16651956
31. Cappellano G, Morandi EM, Rainer J, Grubwieser P, Heinz K, Wolfram D, Bernhard D, Lobenwein S, Pierer G, Ploner C. Human macrophages preferentially infiltrate the superficial adipose tissue. Int J Mol Sci. 2018;19(5):1404. https://doi.org/10.3390/ijms19051404. PMID: 29738484; PMCID: PMC5983635.
32. Dubey NK, Mishra VK, Dubey R, Deng YH, Tsai FC, Deng WP. Revisiting the advances in isolation, characterization and Secretome of adipose-derived stromal/stem cells. Int J Mol Sci. 2018;19(8):2200. https://doi.org/10.3390/ijms19082200. PMID: 30060511; PMCID: PMC6121360.
33. Trivisonno A, Di Rocco G, Cannistra C, Finocchi V, Torres Farr S, Monti M, Toietta G. Harvest of superficial layers of fat with a microcannula and isolation of adipose tissue-derived stromal and vascular cells. Aesthet Surg J. 2014;34(4):601–13. https://doi.org/10.1177/1090820X14528000. Epub 2014 Mar 31. PMID: 24687265.
34. Condé-Green A, de Amorim NF, Pitanguy I. Influence of decantation, washing and centrifugation on adipocyte and mesenchymal stem cell content of aspirated adipose tissue: a comparative study. J Plast Reconstr Aesthet Surg. 2010;63(8):1375–81. https://doi.org/10.1016/j.bjps.2009.07.018. Epub 2009 Aug 12. PMID: 19679523.

35. Di Taranto G, Cicione C, Visconti G, Isgrò MA, Barba M, Di Stasio E, Stigliano E, Bernardini C, Michetti F, Salgarello M, Lattanzi W. Qualitative and quantitative differences of adipose-derived stromal cells from superficial and deep subcutaneous lipoaspirates: a matter of fat. Cytotherapy. 2015;17(8):1076–89. https://doi.org/10.1016/j.jcyt.2015.04.004. Epub 2015 May 19. PMID: 26002819.
36. Aboudib Júnior JH, de Castro CC, Gradel J. Hand rejuvenescence by fat filling. Ann Plast Surg. 1992;28:559–64.
37. Moore JH Jr, Kolaczynski JW, Morales LM, Considine RV, Pietrzkowski Z, Noto PF, Caro JF. Viability of fat obtained by syringe suction lipectomy: effects of local anesthesia with lidocaine. Aesth Plast Surg. 1995;19(4):335–9. https://doi.org/10.1007/BF00451659. PMID: 7484471
38. Rossi M, Roda B, Zia S, Vigliotta I, Zannini C, Alviano F, Bonsi L, Zattoni A, Reschiglian P, Gennai A. Characterization of the tissue and stromal cell components of micro-superficial enhanced fluid fat injection (micro-SEFFI) for facial aging treatment. Aesthet Surg J. 2020;40(6):679–90. https://doi.org/10.1093/asj/sjy142. PMID: 29905790
39. Riesco B, Abascal C, Duarte A, Flores RM, Rouaux G, Sampayo R, Bernardini F, Devoto M. Autologous fat transfer with SEFFI (superficial enhanced fluid fat injection) technique in periocular reconstruction. Orbit. 2018;37(3):191–5. https://doi.org/10.1080/01676830.2017.1383470. Epub 2017 Oct 17. PMID: 29040029.
40. Tonnard PL, Verpaele AM, Zeltzer AA. Augmentation blepharoplasty: a review of 500 consecutive patients. Aesthet Surg J. 2013;33(3):341–52. https://doi.org/10.1177/1090820X13478966. PMID: 23515379
41. Gennai, B. Bovani, M. Colli, F. Melfa, D. Piccolo, R. Russo, Barbara Roda, Andrea Zattoni, Pierluigi Reschiglian and Silvia Zia Comparison of Harvesting and Processing Technique for Adipose Tissue Graft: Evaluation of Cell Viability. Int J Regen Med doi:https://doi.org/10.31487/j.RGM.2021.02.03. 4(2): 2–5.
42. Gennai A, Bernardini FP, Baldessin M, Bovani B, Camporese A, et al. Evaluation of the viability and Phenotipe of adipose derived cells harvested using different harvesting and processing procedures: a pilot study. J stem cells clin pract. 2022;2(1):101.
43. Schiraldi L, Sapino G, Meuli J, Maruccia M, Cherubino M, Raffoul W, di Summa PG. Facial fat grafting (FFG): worth the risk? A systematic review of complications and critical appraisal. J Clin Med. 2022;11(16):4708. https://doi.org/10.3390/jcm11164708. PMID: 36012947; PMCID: PMC9410081.
44. Gennai A, Zia S, Roda B, Maggio A, Bonsi L, Alviano F, Zattoni A, Reschiglian P, Bernardini FP. SEFFI (superficial enhanced fluid fat injection) for aesthetic and clinical regenerative treatments. Glob J Dermatol Venereol. 2020;8:32–40.
45. Gennai A, Baldessin M, Melfa F, Bovani B, Camporese A, Claysset B, Colli M, Diaspro A, Russo R, Strano P, Bollero D, Capparè G, Casadei A, Gallo G, Piccolo D, Salti G, Tesauro P. Guided superficial enhanced fluid fat injection (SEFFI) procedures for facial rejuvenation: an Italian multicenter retrospective case report. Clin Pract. 2023;13(4):924–43. https://doi.org/10.3390/clinpract13040085. PMID: 37623266; PMCID: PMC10453478.

Chapter 2
Facial Aging: A Comprehensive Examination of Time's Impact

2.1 Introduction

Facial aging is a captivating topic that has garnered extensive attention due to its multifactorial nature. The manifestation of age on the face is not merely an alteration of appearance but a testament to the myriads of physiological changes that transpire beneath the skin's surface.

Facial rejuvenation is a multifaceted field that necessitates an intricate understanding of the facial structure, recognizing the significance of addressing various layers for comprehensive results. The face is stratified into five interconnected layers, extending from the neck to the scalp, with each layer harboring unique structures crucial for precise applications and safety during interventions [1].

Remarkably, the acronym "SCALP" delineates these layers: Skin (Layer 1), Connective tissue (Layer 2), Aponeurosis or the musculoaponeurotic layer (SMAS, Layer 3), Loose areolar connective tissue (Layer 4), and Periosteum or deep fascia (Layer 5).

- Skin (Layer 1) exhibits varied characteristics across the face, including differences in pigmentation, thickness, and subcutaneous adherence, playing a pivotal role in facial aging.
- Connective Tissue (Layer 2) encompasses the subcutaneous fat layer, varying in thickness and distribution, impacting the face's contour and volume.
- SMAS (Layer 3), the musculoaponeurotic facial layer, has garnered significant attention, particularly in surgical facelift procedures, where alterations to this layer are fundamental.
- Loose Areolar Connective Tissue (Layer 4) houses the deep fat compartments, separated from the subcutaneous fat by the SMAS, providing a gliding plane for facial muscles and differing in adipocyte size compared to superficial fat [2].
- Periosteum (Layer 5), though directly associated with bone covering in the scalp, manifests as distinct structures across the face, with variations like the deep

D. Piccolo et al., *A.R.T. Autologous Regenerative Therapy in Aesthetic Medicine*, https://doi.org/10.1007/978-3-032-05835-5_2

temporal fascia in the temple, the parotideomasseteric fascia in the lateral mid-face, and continuity with the deep cervical fascia in the neck [3].

Understanding these layers and their interplay is paramount in facial rejuvenation, as addressing only one aspect can lead to suboptimal results. A holistic approach that considers the aging impact on the skeleton, deep and superficial fat compartments, and skin ensures a more balanced, natural, and effective rejuvenation outcome. This comprehensive strategy aligns with the anatomical intricacies of the face, catering to the unique needs of each layer and yielding results that are both aesthetically pleasing and harmoniously integrated.

This chapter provides a detailed exploration of the complexities of facial aging.

2.1.1 The Essence of Facial Aging

The process of facial aging dramatically transforms the harmonious, symmetrical, and balanced attributes of youth. This metamorphosis does not solely diminish physical allure; it also wields significant influence over an individual's self-perception and self-esteem.

2.1.2 Contributing Factors to Facial Aging

Facial aging is a harmonious interplay between changes in the skin, facial skeleton, and soft tissues. Each of these components undergoes distinct alterations as time progresses.

2.1.3 Bone: The Structural Foundation

Facial bones form the foundation that supports the overlying soft tissues. As age progresses, these bones recede and remodel, causing a shift in their structural integrity. This remodeling causes soft tissues to reposition, leading to the typical sagging appearance associated with aging.

This process results in the downward and inward movement of fat pads and muscles, adapting to the altered bony foundation. Craniofacial skeletal remodeling spans from adolescence to middle adulthood and beyond, encompassing changes such as mandibular lengthening or rotation, increase in mandibular angle, and changes in chin projection and jawline. This bone resorption, particularly noticeable in the inferior part of the mandible, and loss of dentition or dentoalveolar regression contribute to an increased mandibular angle and a reduced chin height [4, 5, 58].

Concurrently, the forehead may start to flatten, nasal tip droop, and cheeks hollow, accompanied by deepening of the nasolabial groove and lengthening of the cutaneous upper lip. Midface bone remodeling disrupts the balance between the upper, middle, and lower thirds of the face, leading to increased prominence of bony orbits, enlargement of the pyriform aperture, and shortening of the upper jaw. Additionally, the orbital bone undergoes resorption, causing changes around the eyes and contributing to a rounded appearance and deeper tear troughs [6, 20, 55].

From a lateral view, these changes result in a clockwise rotation of the face relative to the cranial base, leading to a flattening of facial angles. Despite the modest scale of these bony alterations, their impact on the facial appearance is substantial, demonstrating the significant role of bone reabsorption in the aging process [7–9] (Fig. 2.1).

2.1.4 *Fat: The Contour of Youth*

One of the cardinal components in facial aging is fat. Notably:

The identification of facial fat compartments has significantly enhanced our comprehension of midfacial aging, steering the focus from a solely gravitational understanding to a volumetric perspective [10, 11].

The initial discovery of these fat compartments necessitated a variety of methodologies, eventually culminating in a defined anatomical understanding

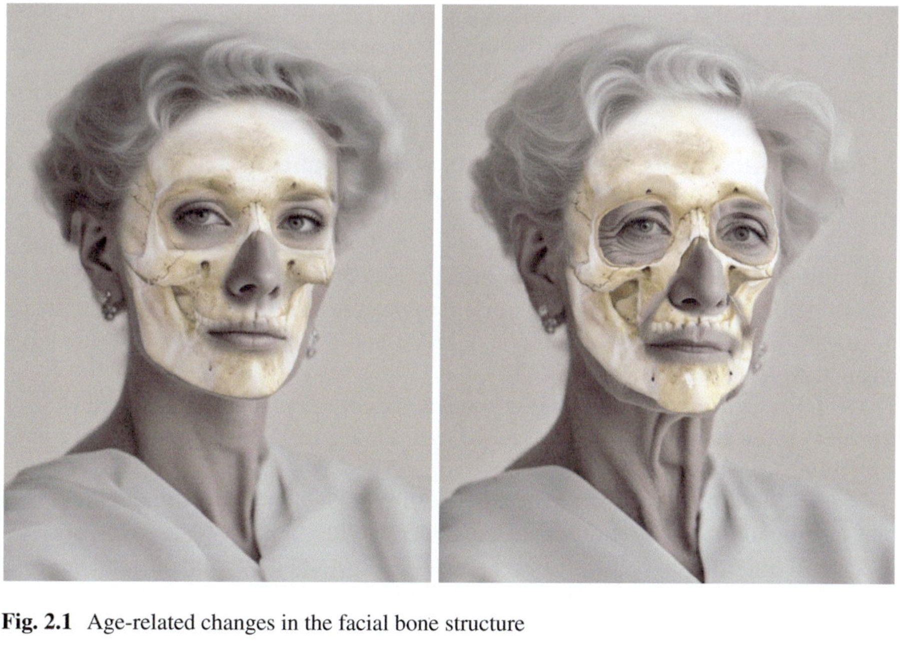

Fig. 2.1 Age-related changes in the facial bone structure

of their structures. This has rejuvenated interest in volumetric facial rejuvenation, with techniques now being guided by the specific augmentation of these compartments. The fat compartments exhibit unique distribution patterns, influenced by the distinct mechanical and biological contexts of the deep and superficial facial fat pads. Traditionally, facial fat was simply categorized into superficial and deep layers, based on their relation to the superficial musculoaponeurotic system and mimetic muscles. These structures provide a clear demarcation between the deep and superficial tissues, though further subdivision into distinct compartments is less straightforward. Rohrich and Rohrich and Pessa's groundbreaking studies in 2007 introduced a reproducible, anatomical method to identify these compartments, utilizing methylene blue dye injections in cadaveric specimens and observing its distribution according to natural septal boundaries [12, 13, 16].

These boundaries serve as a retaining system and form the anatomical foundation for the fat compartments. This technique has been replicated and expanded upon in subsequent research, further detailing the facial fat compartments. In 2012, Gierloff et al. advanced this field by employing CT images to both identify and measure the fat compartments, using iodinated contrast medium for clear visualization on 3D CT images [14].

This method provides a thorough multiplanar and volumetric analysis of the facial fat compartments, adding another layer of precision to our understanding of facial aging and rejuvenation.

These compartments are typically either superficial or deep, in relation to the superficial musculoaponeurotic system. Deep fat remains stationary due to its anchorage to bone, providing shape, supporting overlying fat, and assisting muscle movement. On the contrary, superficial fat is more malleable and reacts to both static and dynamic forces exerted by mimetic muscles.

2.2 Midface Fat Compartments: Classification and Importance

The midface fat compartments are a crucial aspect of facial anatomy, playing a vital role in defining the contours and youthful appearance of the face. These compartments are broadly categorized into superficial and deep layers, each with distinct characteristics and locations (Fig. 2.2).

Deep Fat Compartments: Located beneath the SMAS plane and deep to the facial muscles, these compartments extend to and adhere to the periosteum, providing structural support to the face.

Medial Suborbicularis Oculi Fat (M-SOOF): Situated medially around the eye, it plays a crucial role in the fullness of the lower eyelid.

Lateral Suborbicularis Oculi Fat (L-SOOF): Positioned laterally around the eye, contributing to the contour of the orbital region.

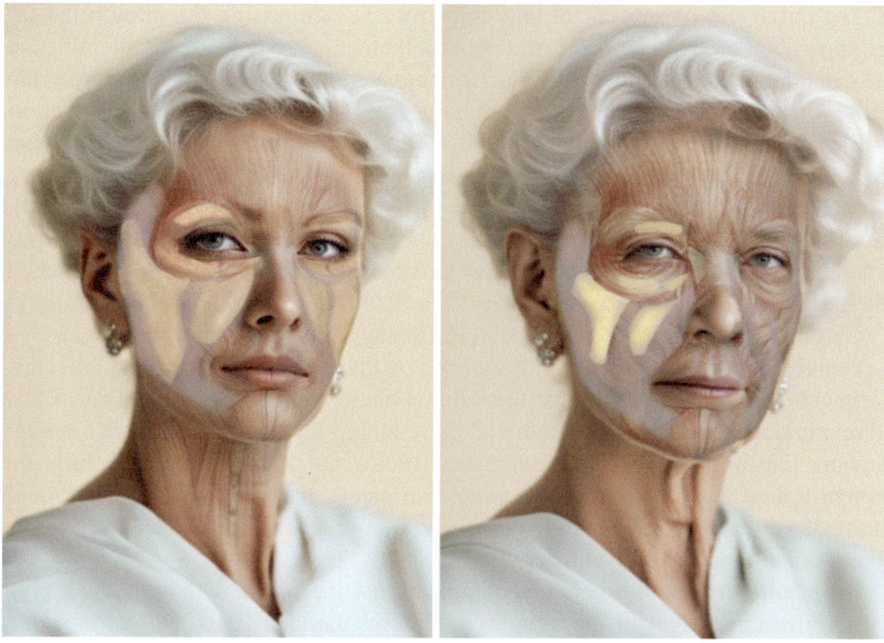

Fig. 2.2 Age-related changes in the deep fat compartments

Deep Medial Cheek Fat (DMCF): Found in the deeper medial cheek area, it provides structural support to the overlying tissues.

Buccal Fat (BF): Located in the central part of the face, it influences the fullness of the cheeks [15].

Superficial Fat Compartments: Positioned between the skin and the Superficial Muscular Aponeurotic System (SMAS), the superficial fat compartments are embedded within a three-dimensional fibrotic network composed of collagen, elastic fibers, muscle fibers, and fat cells.

Infraorbital Fat (IF): Situated just below the eye, this fat compartment helps in maintaining a youthful appearance around the eye area.

Superficial Medial Cheek Fat (SMCF): Located in the medial part of the cheek, contributing to the fullness of the midface.

Nasolabial Fat (NLF): Positioned around the nasolabial fold, playing a role in the appearance of the smile and laugh lines.

Middle Cheek Fat (MCF): Found in the central part of the cheek, it adds volume and contour to the face.

Lateral Temporal-Cheek Fat (LTCF): Extending towards the temporal region, this compartment influences the contour of the side face.

Superior Jowl Fat (SJF): Located above the jawline, it impacts the appearance of the lower face.

Inferior Jowl Fat (IJF): Found below the jawline, contributing to the contour of the jaw and chin area.

The initial trio of superficial fat compartments, comprising the infraorbital fat, the superficial medial cheek fat, and the nasolabial fat, are collectively identified as the malar fat pad. This entity exhibits a triangular configuration, where the base runs parallel to the nasolabial fold, culminating at the apex near the malar elevation. While the malar fat pad maintains a relatively loose connection with the deep plane (SMAS), it exhibits a robust attachment to the skin. During the early years of life, the malar fat pad retains its position due to the presence of numerous fibroelastic fascial septa. These septa traverse the fatty cheek region, originating from the superficial fascia that envelops the muscles responsible for facial expressions, and anchoring into the dermis situated above (Fig. 2.3).

Chronological Progression of Facial Fat Compartment Atrophy in Aging.

Facial architecture is profoundly sculpted by the discrete yet interlinked fat compartments that determine not just the volume but also the shape of the face. These compartments are strategically tethered by retaining ligaments, creating a complex network that influences facial aesthetics. Autologous fat transfer, particularly the replenishment of volume in the deep compartments, has become a cornerstone in the realm of facial rejuvenation. This technique is utilized judiciously at various stages of aging to restore youthful vitality.

In the third decade of life, the fat compartments are robust, with individuals typically showing no signs of the atrophy that marks the later years. However, as one progresses into the fourth and fifth decades, changes become more evident. Atrophy begins to manifest in the superficial lateral and medial compartments, as well as in

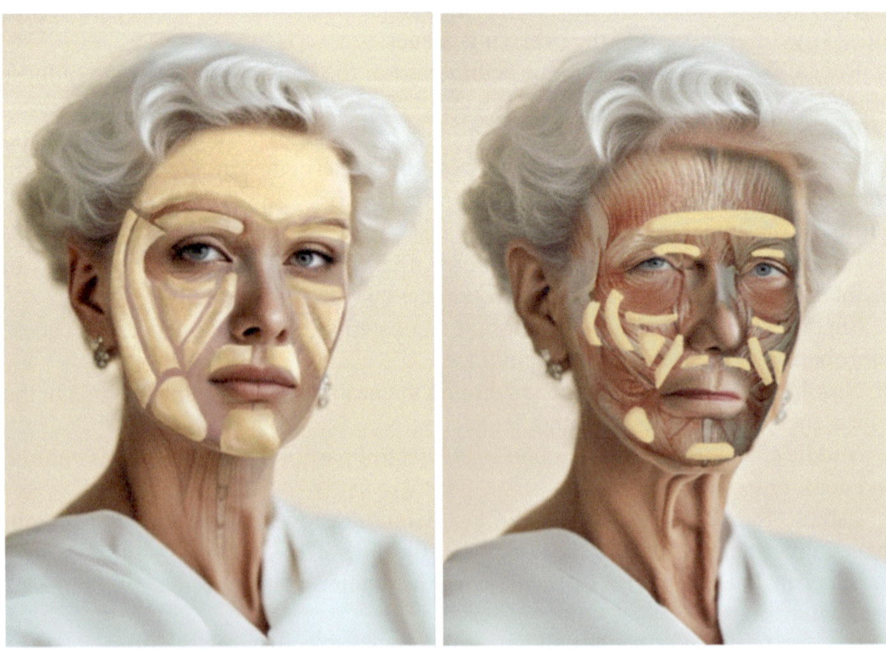

Fig. 2.3 Age-related changes in the superficial fat compartments

the deep medial cheek and the Suborbicularis Oculi Fat (SOOF). These changes contribute to the shifting landscape of the face, as volume diminishes, and the youthful contour begins to wane.

Transitioning into the sixth and seventh decades, the aging process extends, now encompassing superficial malar and deep lateral cheek atrophy. This further alteration in the facial fat compartments accentuates the signs of aging with more pronounced hollowing and sagging.

Beyond the 70th year, the landscape of aging encompasses even the lateral temporal and forehead compartments, areas previously resilient, now succumbing to the relentless progression of time.

It is imperative, however, to acknowledge that these observations are not absolute. The rate and pattern of fat compartment atrophy are highly individualized, making it difficult to standardize against specific ages. The schematic presented here is distilled from recent research, serving as a general framework to understand the chronological progression of facial fat compartment aging, while still recognizing the vast diversity found within the human population [11].

2.2.1 *Muscles: The Dynamic Canvas*

In the intricacies of facial aging, muscular transformations can play a role, manifesting through a spectrum of changes in muscle size, tone, elasticity, and position. As individuals age, certain facial muscles are prone to atrophy, leading to a reduction in volume and sagging of the overlying skin, while others may undergo hypertrophy due to repetitive facial expressions, potentially altering the facial contour, as observed in the enlargement of the masseter muscle [16].

The diminished muscle tone and loss of elasticity can contribute to the decreased support for the skin, resulting in the formation of wrinkles and a sagging appearance [17].

Furthermore, an imbalance in muscle activity becomes apparent; hyperactivity in specific muscles leads to pronounced expression lines, while underactivity results in volume loss and a lackluster appearance [18].

These muscular changes are intricately linked with alterations in neurotransmitter levels, affecting muscle contraction and contributing to an overall stiffness or slackness of the facial muscles [19].

The repositioning of muscles due to weakened ligaments and connective tissues further influences the facial contours, often leading to a tired or sagging appearance. Additionally, the interaction between muscular transformations and other aging components such as skin elasticity, fat distribution, and bone structure create a complex network, collectively contributing to the aging phenotype.

Interventions such as the use of neuromodulators can temporarily alleviate hyperactivity of muscles and smooth out wrinkles. On the other hand, strategies aimed at maintaining muscle tone, including specific facial exercises and massage, may offer benefits, though their efficacy varies across individuals [20].

It is incumbent to report that some authors as Gosain AKet al. proved with MRI analysis that there are no significant changes for both mimetic muscles, there were no significant differences between young and old subjects in muscle length, thickness, or volume from muscle origin to nasolabial fold. In addition, there were no significant differences between age groups in fatty infiltration of the muscles [21].

2.2.2 *Skin: The External Reflection of Aging*

The trajectory of skin aging is shaped by a combination of internal and external factors:

- Intrinsic Aging: Governed by genetic predisposition and the inevitable passage of time, this type of aging manifests differently across individuals. Its hallmarks include the subtle emergence of fine lines and a uniform scattering of pigmentation.
- Extrinsic Aging: Here, external elements such as UV radiation, tobacco use, and environmental pollutants including airborne contaminants are key contributors. This aging variant is characterized by more pronounced wrinkles, uneven skin tone, a rough texture, skin dryness, visible blood vessels (telangiectasia), and a yellowish discoloration of the skin [22–24].

Central to the skin's ability to maintain its shape and resilience is the dermis, rich in collagen, elastin, and glycosaminoglycans:

- Collagen: Over time, there is a noticeable reduction in collagen production coupled with an increase in its degradation, leading to diminished skin firmness.
- Elastin: Production of elastin significantly decreases in the 40–50 age bracket, and the protein undergoes degradation. This, alongside the breakdown of the elastic fiber network, impairs the skin's elasticity.
- Glycosaminoglycans: These molecules are vital for skin hydration, but their degradation contributes to skin dryness, accelerating the aging process [25, 26].

Extrinsically aged skin additionally displays disarray in collagen fiber arrangement, compromising the dermis's structural stability. There is an initial increase in the thickness and quantity of disordered elastic fibers, but this is followed by their degradation, resulting in a reduction of tissue elasticity. The outcome is the formation of permanent lines and temporary creases as the skin yields to the ongoing pull from facial muscles.

Crucially, the dermis's composition of collagen, elastin, and glycosaminoglycans plays a significant role in the skin's resilience and ability to counteract various stresses. In its youthful and healthy state, the skin benefits from a network of elastic microfibrils attached to an elastin core within the extracellular matrix, endowing the skin with flexibility and the ability to revert to its initial state post-stretching. However, the intrinsic aging process leads to a depletion and deterioration of these essential dermal components, resulting in thinner, weaker skin. The rate of collagen degradation outpaces its production, and from the ages of 40 to 50 onwards, there is a dramatic decrease in elastin synthesis. This results in a compromised elastic fiber network and reduced skin hydration due to the breakdown of glycosaminoglycans [27].

The influence of extrinsic aging on these dermal components intensifies the aging signs seen in intrinsic aging (Fig. 2.4). Apart from the heightened collagen breakdown and diminished production, collagen fibers become disordered, undermining the dermis's structural integrity. Glycosaminoglycans, although increasing in quantity, accumulate in disordered clusters and lose their ability to regulate hydration, giving the skin a leathery texture. The most severe effect of extrinsic aging is solar elastosis, marked by an initial proliferation of abnormal elastic fibers, followed by their subsequent degradation. This leads to a loss of skin resilience, resulting in the appearance of static wrinkles and dynamic folds as the skin succumbs to the constant pull from mimetic muscles. On a molecular level, an increase in matrix metalloproteinase expression and an upsurge in reactive oxygen species activity—largely driven by UV exposure—gradually degrade the dermal matrix. Aging is also associated with a gradual reduction in cell number and function, including melanocytes and Langerhans cells, as well as a decline in hormone levels that influence skin physiology [28].

2.2.3 *Molecular Mechanisms*

Skin aging is a multifactorial process influenced by genetic, metabolic, and environmental factors, manifesting through a range of cellular and molecular changes [29].

Fig. 2.4 Age-related changes in the skin

Cellular senescence, DNA damage, oxidative stress, and mitochondrial dysfunction collectively lead to decreased cell regeneration, energy production, and repair capabilities, resulting in thinner, less elastic skin.

Chronic inflammation and glycation further contribute to the degradation of collagen and elastin, key structural proteins, leading to wrinkles and sagging [30–32].

Hormonal fluctuations, particularly in estrogen and testosterone, along with decreased sebum production, exacerbate skin dryness and contribute to the loss of skin thickness and elasticity [33].

The shortening of telomeres with each cell division also plays a crucial role, as it triggers cellular senescence or death, contributing to the visible signs of aging [34].

Understanding these intricate molecular mechanisms is pivotal for the development of targeted antiaging interventions and therapeutic strategies.

2.3 Role of Autologous Regenerative Therapy in Aesthetic (ARTA) for Facial Rejuvenation

Facial aging is a multifaceted process that encompasses all facial structures from the innermost layers, such as the bones, to the outermost layers, like the skin. The transformations that these structures undergo are influenced by various factors including age, genetics, and lifestyle habits. In order to address these changes effectively, a comprehensive facial rejuvenation plan must be established. This plan should aim to meticulously strike a balance between restoring volume to the tissues, rebalancing muscle activity, tightening the skin, and enhancing the skin's overall quality.

Adopting a holistic and standardized approach is paramount, as it represents the gold standard in modern medical treatments for facial rejuvenation. In this context, Autologous Regenerative Therapy in Aesthetic (ARTA), particularly when utilizing adipose tissue, takes on a pivotal role due to its intrinsic properties that harness the body's natural healing and regenerative capabilities [35, 36].

Adipose-derived stem cells (ADSCs) found within the adipose tissue are central to this process, as they exhibit a remarkable capacity for tissue repair and regeneration [37].

They possess the potential to differentiate into a variety of cell types, as well as to release growth factors and cytokines that are instrumental in stimulating collagen production, enhancing skin elasticity, and promoting the formation of new blood vessels. Moreover, the anti-inflammatory properties of ADSCs contribute to reducing post-procedural redness and swelling, ensuring a smoother and quicker recovery [38, 39].

Unique to this therapy is its ability to provide long-lasting and sustainable results, offering more than just a temporary solution and actively participating in the continuous rejuvenation of facial tissues. Furthermore, ARTA's autologous nature ensures a commendable safety profile, minimizing the risk of allergic reactions or tissue rejection, solidifying its position as a natural and effective choice for facial rejuvenation.

Recognizing the central role of ARTA in facial rejuvenation, it becomes clear that it should be integrated with other aesthetic procedures to achieve harmonious and natural results [40].

The principal aim of this book is to furnish medical professionals with a precise, detailed, and standardized guideline for approaching facial rejuvenation, ensuring the best possible outcomes for patients seeking to rejuvenate and revitalize their facial appearance.

2.4 Microfat and Nanofat: Dual Pillars of Adipose-Based Facial Rejuvenation

The facial aging is a process that involves the whole structure of the face, bone, muscles, deep and superficial fat compartments, and skin. The modification in fat compartments (deep and superficial) plays a fundamental role in facial aging [10, 11] (Fig. 2.5).

The reduction in volume of the deep fat compartments leads to a phenomenon known as "pseudoptosis" of the superficial fat [62].

This condition arises when the underlying deep fat pads lose volume and no longer support the fat above, leading to a downward migration of the superficial fat. These insights have significantly enhanced the appreciation of facial fat grafting as a potent technique for facial rejuvenation, underscoring its importance in restoring youthful contours to the face.

Fig. 2.5 The patient's assessment should be performed from inside out

The role of the Autologous Regenerative Therapy Aesthetic (ARTA) is to restore the volume of the deep and superficial fat compartments and regenerate the skin. In the light of these evidences, ARTA plays a fundamental role in the facial rejuvenation strategy.

Microfat and nanofat represent transformative approaches in the realm of ARTA, harnessing the inherent potential of adipose tissue for diverse clinical applications [41–43].

Microfat, characterized by its larger adipose clusters, plays a pivotal role in volumizing and contouring, addressing tissue atrophy and volume loss that occur due to aging or other factors. Its ability to provide immediate mechanical support and induce long-term tissue integration makes it a preferred choice for restoring facial volume, smoothing contours, and enhancing overall aesthetic appearance.

On the other hand, nanofat, with its fine, cellular composition, stands out for its regenerative and rejuvenating capabilities [44–46].

Rich in adipose-derived stem cells, growth factors, and a plethora of bioactive molecules, nanofat contributes to skin rejuvenation, enhancing texture, elasticity, and tone. It has shown promise in treating fine lines, scars, and other signs of skin aging, leveraging its potent regenerative properties to stimulate tissue repair and renewal at a cellular level.

Together, microfat and nanofat encapsulate the dual aspects of adipose-based therapies: volumizing and regenerative. Their purposeful use, individually or in combination, enables clinicians to adopt a holistic approach to facial rejuvenation, tailoring interventions to the unique needs and aesthetic goals of each patient. Through this precision and versatility, microfat and nanofat play integral roles in advancing the field of aesthetic medicine, offering solutions that are as effective as they are natural and harmonious with the body's own tissues.

This approach ensures a balanced and comprehensive strategy to facial aging, addressing not just the superficial signs of aging but also the underlying structural changes, to achieve outcomes that are both visually pleasing and enduring.

2.5 Guided SEFFI: The Right Fluidity for the Right Purpose

In the preceding chapter, we described the guided Superficial Enhanced Fluid Fat Injection (SEFFI), a pivotal guided technique devised to safely, simply, and effectively obtain adipose tissue for Autologous Regenerative Therapy Aesthetic (ARTA), while ensuring standardization even in instances where the practitioner may not possess extensive liposuction expertise. The guided SEFFI cannula is a derivation of the original SEFFI technique [47, 48], where a microcannula with 800-micron side-port holes is utilized to meticulously harvest superficial adipose tissue (SAT). This deliberate selection of cluster dimensions during the collection phase allows for the acquisition of microfat, bypassing the need for substantial manipulations often involving filters, meshes, spheres, and other devices.

The importance of attaining adipose tissue of varying fluidity for ARTA (microfat to nanofat) applications on the face was a focal point of our discussion [46, 49] (Fig. 2.6).

In response to this, we have established a standardized procedure to modify the fluidity of the tissue acquired through the guided SEFFI method. This procedure incorporates the use of two 10 ml syringes interconnected via a simple, filter-free connector. By gently transferring the tissue back and forth between the syringes, one can incrementally enhance the fluidity of the tissue and reduce the dimensions of the tissue clusters.

The progression follows a structured scheme: starting with 800-micron microfat, 2–3 passages yield tissue of 600 microns; 5–6 passages result in 500 microns; 10–11 passages produce 400 microns; and eventually, 20–30 passages culminate in nanofat of 200 microns (Fig. 2.7).

Adhering to this straightforward protocol enables the practitioner to generate adipose tissue of diverse fluidities, tailor-made for injection into varying layers of the face tissues, addressing distinct aesthetic objectives.

The underlying principle is succinct yet profound: larger clusters result in less fluid tissue, favoring volumizing effects, whereas smaller clusters yield more fluid tissue, enhancing regenerative outcomes [46].

By mastering this technique and understanding the nuanced relationship between cluster size, tissue fluidity, and their respective aesthetic impacts, practitioners are empowered to offer nuanced, personalized ARTA treatments, ensuring optimal results and patient satisfaction.

Building upon the foundational knowledge regarding the guided Superficial Enhanced Fluid Fat Injection (SEFFI) for Autologous Regenerative Therapy Aesthetic (ARTA), it is crucial to highlight the significant advantages that this

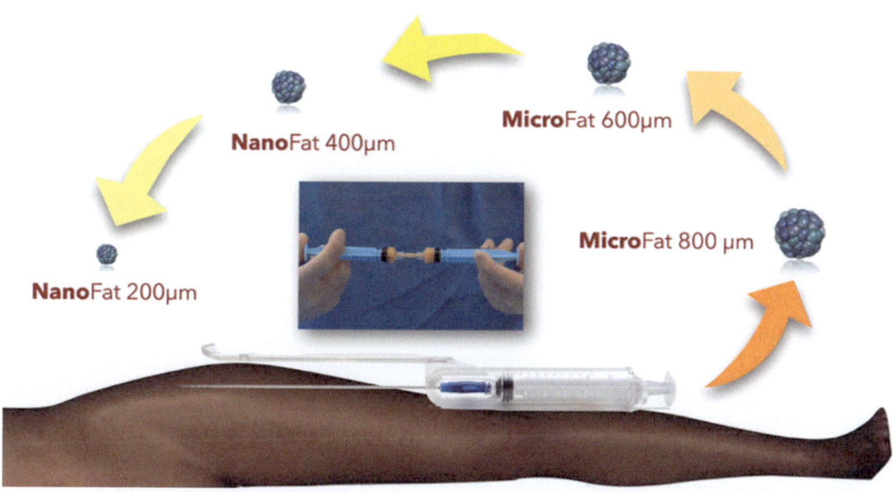

Fig. 2.6 Simply passaging the tissue from one syringe to another is possible to reduce the clusters dimension form microfat to nanofat

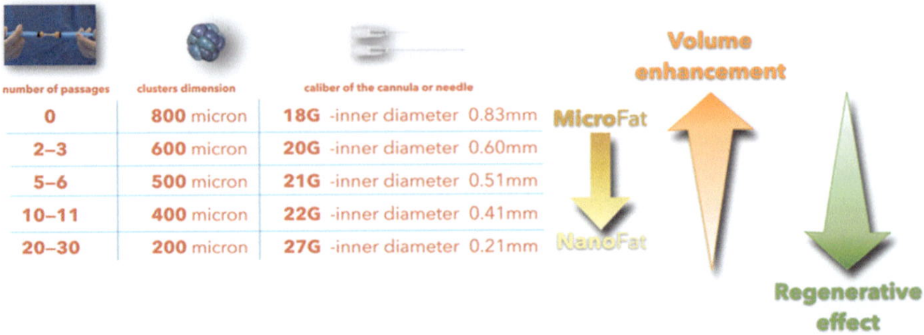

Fig. 2.7 Based on the number of times the tissue is passed from one syringe to another, the size of the clusters is reduced—transitioning from microfat to nanofat. As microfat is transformed into nanofat, its volumizing capacity decreases while its regenerative potential increases

innovative technique brings to the table. One of the benefits of utilizing the guided SEFFI method is its independence from additional medical devices for reducing tissue cluster dimensions; there is no necessity for filters, meshes, spheres, or similar apparatus. This not only simplifies the procedure but also ensures that there is minimal to no waste of tissue, contributing to the overall efficiency of the process [47–49].

The lack of extensive manipulation means that the tissue collection can be performed in smaller quantities, making the procedure faster, less traumatic, and inherently safer for the patient. This aspect is particularly critical, as it minimizes the potential for complications and expedites the recovery period, enhancing the patient's experience and satisfaction with the treatment.

Another significant advantage of the guided SEFFI technique lies in its capacity to facilitate the achievement of various degrees of tissue fluidity through simple passages of the adipose tissue from one syringe to another. This versatility in manipulating the tissue's consistency allows for a high level of customization, enabling practitioners to tailor the procedure precisely to meet the unique needs and aesthetic goals of each individual patient.

Moreover, the gentle handling of the adipose tissue intrinsic to the guided SEFFI procedure plays a vital role in preserving the viability of the cells, maintaining their stemness, and promoting an optimal growth rate. This is crucial for ensuring the long-term success of the ARTA treatment, as it directly impacts the regenerative capacity of the transplanted adipose tissue, contributing to more effective and enduring aesthetic outcomes [50, 51].

In summary, the guided SEFFI technique for ARTA stands out for its simplicity, efficiency, and capability to provide a customizable and safe treatment option, all while ensuring the highest possible viability and functionality of the adipose tissue. These advantages collectively position the guided SEFFI method as an invaluable tool in the realm of regenerative aesthetics, offering patients a reliable and effective solution to meet their aesthetic and regenerative needs. The guided SEFFI involves the use of a microcannula with 800 micron side-port holes and then the tissue obtained without any manipulation.

2.6 Regenerative Facial Rejuvenation Strategy: From the Inside Out

Facial aging is a complex and multifactorial process. It represents the cumulative impact of time on the skin, underlying soft tissues, and the deeper structural components of the face. This aging process is characterized by a combination of factors: changes in skin texture, reduction in facial volume, bone resorption, loss of tissue elasticity, and the effects of gravity (Fig. 2.8). Together, these factors contribute to the visible signs of aging [52].

In order to correct these factors, it is important to evaluate the face "from inside out" using the right fluidity of tissue, from micro to nanofat, considering that moving from micro to nanofat, the volumizing effect decreases and the regenerative effect increases (Fig. 2.9).

To comprehensively assess and address the morphological effects of aging, it is effective to divide the face into three distinct regions: the upper, middle, and lower thirds.

Fig. 2.8 Age-related facial compartments changes

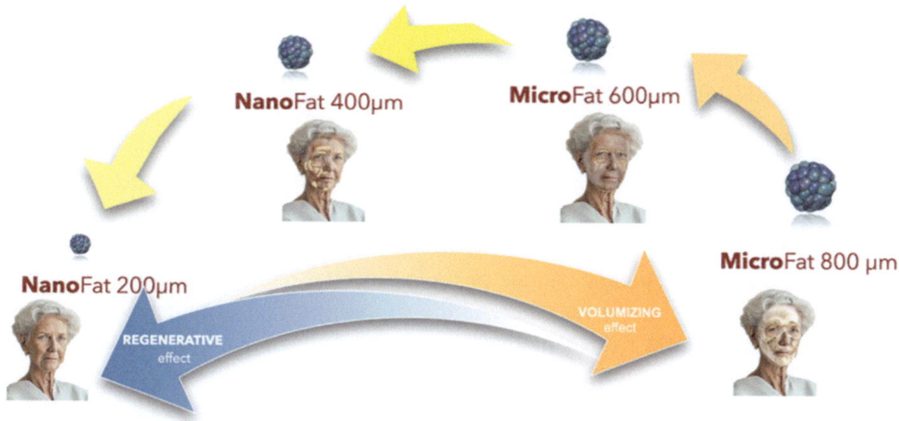

Fig. 2.9 The treatment of the compartments involved in facial aging, from the inside out, requires varying tissue fluidity—from microfat to nanofat. As microfat is transformed into nanofat, its volumizing capacity decreases while its regenerative potential increases

2.7 Upper Third

In the upper third of the face, aging typically manifests as a loss of subcutaneous fullness in the forehead, brow, temple, and upper eyelid regions. This loss accentuates the underlying bone structure and leads to visible changes. In the forehead and glabellar areas, the most common signs of aging are the formation of wrinkles and an increase in the nasofrontal angle, which results in a flattened appearance and decreased projection. The temporal region also undergoes significant changes, primarily characterized by a reduction in the thickness of superficial fat pads, leading to a narrowed bitemporal width and a more scalloped appearance [53].

An MRI study confirmed that the soft tissue thickness in the temporal region decreases significantly throughout adulthood [53].

Additionally, the lateral portion of the eyebrows may sag, primarily because the lateral edge of the frontalis muscle does not have a direct point of insertion at the temporal fusion line. This anatomical peculiarity results in less structural support for the lateral eyebrow [9].

Furthermore, the loss of temporal support to the lateral brow, combined with a decrease in upper eyelid fullness, creates an impression of brow ptosis, where the eyebrows appear to descend to or below the superior orbital rim [54].

The weakening of the orbital septum can lead to the protrusion of intraorbital fat, giving the orbit a more bony appearance. The cumulative effect of these changes, including fixed glabellar frown lines, transverse forehead furrows, temporal hollowing, and a skeletonized supraorbital rim, contributes to the overall impression of aging in the upper face.

2.8 Middle Third

The middle third of the face often shows the earliest signs of aging, particularly in the periorbital area. Changes in skin color and texture are common, as are variations in the theories explaining the development of under-eye bags. These bags may result from either a loss of structural support or the stretching of Lockwood's ligament, leading to a descent of the eye globe and a forward push of the intraorbital fat. Concurrently, the bony recession of the orbital rim enlarges the orbital aperture in both width and area. The appearance of the periorbital region is also greatly influenced by the volume of the fat pads. As the supraorbital fat diminishes, the upper eyelid arc's peak shifts from a medial to a more central position, resulting in the characteristic A-frame deformity [9, 55, 56].

The lower eyelids also undergo changes, primarily descending and leading to fat accumulation and scleral show. The forward herniation of infraorbital fat occurs as the inferior orbital rim recedes and the skeletal support becomes inadequate to maintain the attenuated soft tissues. Moreover, the loss of volume in the deep cheek fat compartments exerts additional traction on the lower eyelid, contributing to the scleral show. The spectrum of age-related changes in the upper and lower eyelids includes sagginess, under-eye bags, and a hollow appearance. The upper eyelids descend, elongating the upper lid length, disrupting the alignment of the upper eyelid fold and upper lid margin, altering the projection of the tarsal region, and increasing the visibility of the pretarsal region. This may be exacerbated by a loss of elasticity and deflation of the upper eyelid's subcutaneous tissue. Crow's feet lines, primarily resulting from repetitive contraction of the lateral orbicularis oculi muscles when smiling, become more pronounced with fat loss and skin thinning. The tear trough refers to the hollow between the eye area and the upper medial cheek, extending slightly below the orbital rim. It forms part of the nasojugal groove, where the orbicularis muscle is firmly attached to the underlying bone. The lid-cheek junction, which comprises the nasojugal groove section extending below the orbital rim, becomes more prominent with age due to deflation of soft tissue, fat atrophy, and descent. Furthermore, tear trough severity intensifies with the deflation of the deep medial cheek fat, together with the weakening of the zygomatic–cutaneous and orbital ligaments: [7, 10, 20, 28, 57–62].

2.9 Lower Third

In the lower third of the face, aging is marked by increased skin laxity, depletion of malar and perioral fat deposits, and alveolar bone resorption, leading to a loss of definition in the jawline. The mandibular fullness dissipates, and the chin undergoes changes in volume and projection. Bony changes in the perioral region, such as mandible shrinkage and dentoalveolar regression, lead to a structural shift in the mandible's shape. Subcutaneous fullness recedes, revealing jowl fat and leading to the development of a "turkey neck" deformity [9, 52].

The chin zone undergoes changes in height, width, and projection, with gender-specific differences in mandibular rotation affecting chin projection. Women tend to have decreased chin projection, as their mandibles rotate inferiorly and backwards toward the skull, whereas men tend to have increased chin projection, as their mandibles undergo more forward rotation [63].

The loss of definition in the jawline is further exacerbated by increased skin laxity and a decrease in the volume of both the superior and inferior jowl fat compartments, which extend downward from the nasolabial fold. The sagging of jowl fat is also influenced by the weakening of the mandibular septum, the structural component responsible for maintaining the position of these fat compartments. As this septum weakens, the support for the jowl fat diminishes, leading to more noticeable sagging and a pronounced jowly appearance [64].

In summary, facial aging is a detailed and nuanced process that affects each region of the face in distinct ways. Understanding these changes is crucial for effective rejuvenation strategies and treatments.

2.10 Case Evaluation and Planning Strategies for Facial Regenerative Rejuvenation Treatment

2.10.1 Introduction

In this chapter, we explore the intricacies of facial rejuvenation, focusing on a patient-centered approach that begins with detailed photographic documentation and progresses through a comprehensive evaluation to a tailored treatment strategy.

2.11 Initial Photographic Documentation

The journey begins with capturing the patient's facial features in their most natural state. We ensure the hair is neatly tied back, the face is free of makeup, and the expression is relaxed. Photographs from various angles—frontal, three-quarter, and both side profiles—provide a foundational understanding of the facial structure and the signs of aging.

2.12 Patient Evaluation in Standing Position

Evaluating the patient in a standing position offers insights into how gravity affects facial tissues. This perspective is invaluable for appreciating the natural descent and displacement of facial features, influencing our treatment plan.

2.13 Identifying the Facets of Aging

The heart of our approach lies in a thorough evaluation to pinpoint the primary factors contributing to facial aging. This involves not only a clinical assessment of skin tone, texture, and volume but also a discussion with the patient about their perceptions of their aging areas. This dual approach ensures a holistic understanding of the patient's needs and expectations.

2.14 Patient's Assessment and Strategic Planning

A precise patient's assessment is crucial. This hands-on assessment, combined with the photographic analysis, guides our planning. We chart out the specific areas to be treated, considering various aspects such as the treatment planes (bone, deep and superficial adipose compartments, skin), the materials to be used, their consistency, and quantities (Fig. 2.10).

2.14.1 Treatment Strategy

2.14.1.1 Bone Support with CaHA

Calcium hydroxylapatite (CaHA) is our first choice for enhancing bone support. Its compatibility with the body's natural processes makes it a safe and effective option. We generally recommend diluting 1.5 ml of CaHA with 0.5 ml of saline solution [65, 66].

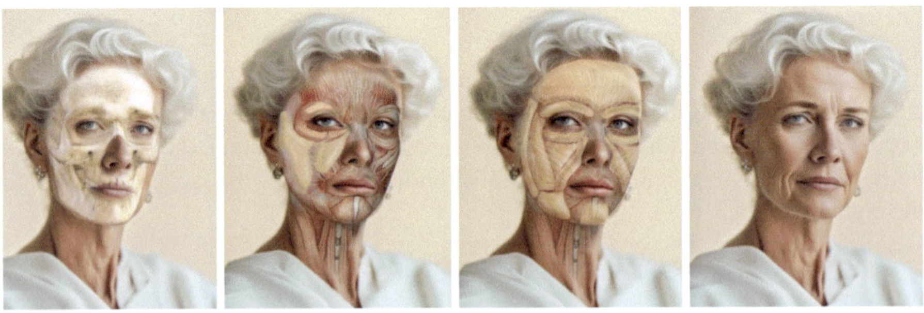

Material	ml	fluidity	total

Material	ml	fluidity	total

Material	ml	fluidity	total

Material	ml	fluidity	total

Fig. 2.10 Pretreatment chart for patient's assessment

2.14.1.2 Volume Restoration with Guided SEFFI

For deep fat compartments, we use guided SEFFI microfat (800–600 micron), which is crucial for restoring significant volume losses.

The superficial compartments are treated with a more refined version of guided SEFFI microfat (600–400 micron), targeting surface-level volume restoration.

2.14.1.3 Skin Rejuvenation

Nanofat (200 micron), applied via the guided SEFFI technique, rejuvenates the skin, enhancing texture and vitality.

2.14.1.4 Fluidity and Regenerative Effects

The fluidity of the guided SEFFI tissue is key. Higher fluidity lessens the volumizing effect but enhances regenerative capabilities. Conversely, lower fluidity provides more volume enhancement.

2.15 Age-Specific Planning and Technical Approaches for Regenerative Facial Rejuvenation

2.15.1 Introduction

Facial rejuvenation, a crucial aspect of aesthetic medicine, entails a multifaceted approach to counteract the aging process. This process impacts the skin, subcutaneous tissues, and structural aspects of the face. The strategy for rejuvenation must be tailored to the patient's age and the specific characteristics of their aging process, taking into account skin texture changes, facial volume loss, bone resorption, tissue elasticity decline, and gravity's effects.

2.15.2 Age-Related Changes and Strategies

In the 30 s: The onset of aging signs is often subtle, marked by a general appearance of fatigue or a "sad" facial expression, rather than specific localized changes. The focus here is on early intervention and preventive measures, emphasizing treatments that maintain tissue vitality and slow aging progression.

From 35 to 45 Years (Fig. 2.11): This period witnesses noticeable changes, particularly in the periocular region:

Fig. 2.11 35- to 45-year-old patient

2.15.3 Regenerative Treatment Planning

The evaluation of the patient should proceed from the inside out.

Firstly, does she require bone support? At her age, extensive bone support is seldom necessary. However, sometimes support is beneficial, particularly in the malar area close to the orbit.

My plan includes injecting multiple micro boluses of CaHA (1.5 ml of CaHA diluted in 0.5 ml saline solution) into the medial malar area, gradually decreasing the volume towards the zygomatic area.

The injection technique involves a 1.5 ml luer lock syringe with a 27G needle, performing deep injections above the periosteum (Fig. 2.12).

In this area is mandatory to mark on the skin the infraorbital nerve in order to inject filler away from this anatomical structure:

- The Infraorbital Nerve: It is a sensory nerve and it is a continuation of the maxillary nerve (V2), which itself branches from the trigeminal nerve. The infraorbital nerve runs forward in the floor of the orbit in the infraorbital groove, enters the infraorbital canal, and emerges onto the face through the infraorbital foramen that is located just below the lower margin of the orbit, and about 1 cm below the infraorbital margin, the nerve exits the skull here.

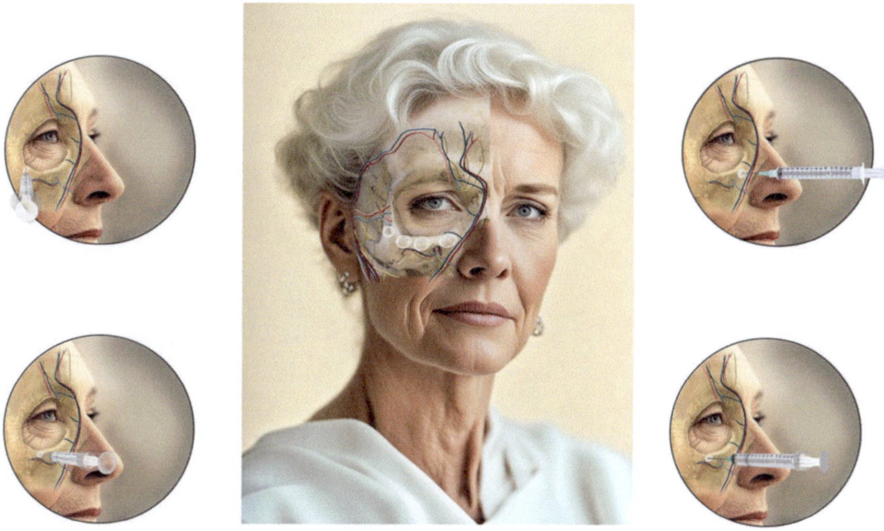

Fig. 2.12 CaHA deep injections above the periosteum

Are deep fat compartments enhancement needed? Typically, at this age, enhancing deep compartments is unnecessary, except for the deep fat compartment of the inferior eyelid. Sometimes, the inferior sulcus appears slightly hollow and should be treated. This involves injecting 1–1.5 ml of tissue at 500 microns (5–6 passages of fluidification) beneath the orbicularis muscle using a 21G cannula. Prior to injection, administering a small amount of local anesthesia with adrenaline (1:100,000) is advisable. The entry point is the lateral inferior border of the orbital frame; the cannula is inserted perpendicularly, touching the bone, then rotated 90 degrees, and slid under the orbicularis muscle (Fig. 2.13) The injection is linear and retrograde.

Does she require superficial fat compartment enhancement? At her age, these compartments are typically impacted by aging.

In the upper third, treatment of superficial compartments includes the tail of the eyebrow and the temporal area. Here, I plan to inject 2–4 ml of tissue at 600 or 500 microns (2–3 passages of fluidification), depending on skin thickness, using a 20–21G cannula. The injections in all areas are performed linearly and retrograde, following an initial administration of a small amount of local anesthesia with adrenaline (1:100,000). The injection in the eyebrow is subcutaneous, while in the temporal area, it is above the major fascia of the temporal muscle. In the temporal area, the entry point is made caudal to the temporal crest; the cannula slides over the fascia, starting the injection from the caudal part of the eyebrow and gradually moving downwards to the temporal area, up to the orbital frame and zygomatic arch.

In the middle third, the treatment targets the malar fat pad and the nasolabial fold.

Enhancing the superficial fat pad in the malar area is key to correcting mid-face deflation, which causes tissue descent and nasolabial fold prominence.

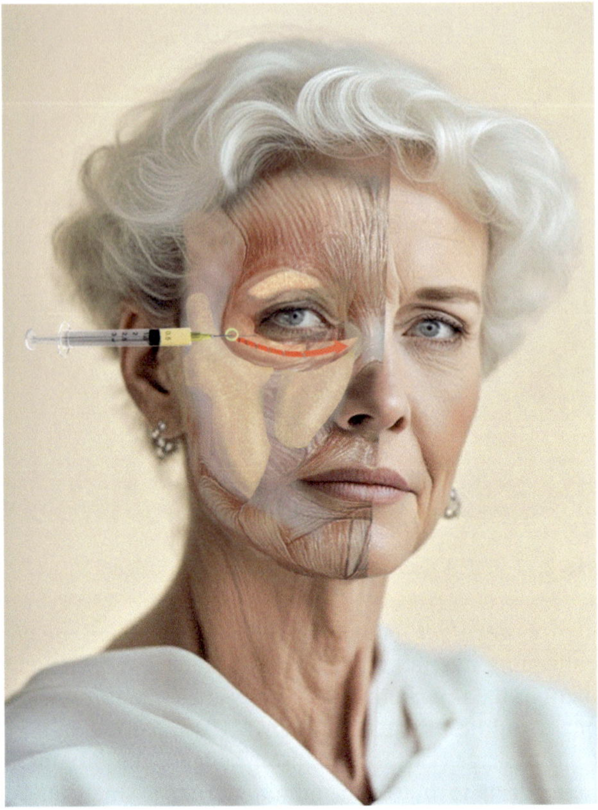

Fig. 2.13 Tissue injection in the inferior border of the orbital frame

An anesthetic block of the infraorbital nerve is often considered. Typically, I inject 3–5 ml of tissue at 600 microns (2–3 passages of fluidification) using a 20G cannula.

The entry points can be either in the inferior or superior lateral malar area; the injection is in the superficial subcutaneous layer and performed using a linear retrograde technique.

In the lower third of the face, treatments are focused on the perioral region and jawline. I typically address the "Marionette" area by injecting 2 to 2.5 milliliters of tissue at 600 microns (2–3 passages of fluidification) using a 20G cannula, with my preferred entry point being the center of the chin. For the jawline and preauricular area, I favor the mandibular angle as the entry point. Here, I administer 2 to 3 milliliters of tissue at 600 microns using a 20G cannula. In the preauricular area, the usual approach involves injecting 2 to 2.5 milliliters of tissue at 500 to 600 microns, utilizing either a 20G or 21G cannula (Fig. 2.14).

Between 45 and 55 Years (Fig. 2.15): This age range sees a deepening of facial lines and more pronounced changes:

- Increased laxity in the upper eyelid and further descent of the eyebrows

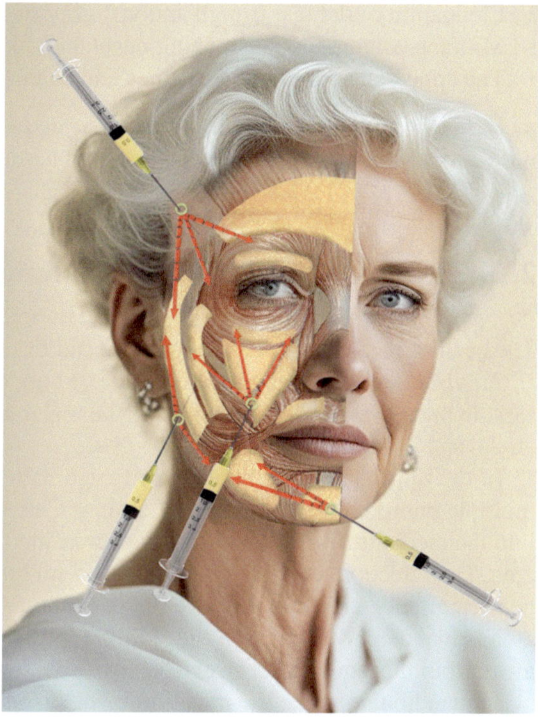

Fig. 2.14 Tissue injection of the superficial fat compartments

Fig. 2.15 45- to 55-year-old patient

- Elongation of the tear trough and visibility of the infraorbital rim
- Midface volume loss, deepening nasolabial folds, and thinning lips
- The appearance of marionette lines and changes in jawline definition

Here, the approach should intensify to counteract these pronounced changes, with a focus on volume restoration and skin texture improvement.

2.16 Regenerative Treatment Planning

The evaluation of the patient should proceed from inside out.

Firstly, does she require bone support? At her age, a bone support is often necessary, in particular, in the upper part of the malar and zygomatic area, in mandibular angle, and in the chin.

In the upper part of the malar and zygomatic area, my plan includes injecting multiple micro boluses of CaHA (1.5 ml of CaHA diluted in 0.5 ml saline solution) into the medial malar area, gradually decreasing the volume towards the zygomatic area. The injection technique involves a 1.5 ml luer lock syringe with a 27G needle, performing deep injections above the periosteum. As previously mentioned, in this area is mandatory to mark on the skin the infraorbital nerve in order to inject filler away from this anatomical structure.

In mandibular angle, the injection involves one to three micro boluses of CaHA (1.5 ml of CaHA diluted in 0.5 ml saline solution), and injection is performed with 1.5 ml luer lock syringe with a 27G needle injections above the periosteum.

In the chin, the injection involves one to three micro boluses of CaHA (1.5 ml of CaHA diluted in 0.5 ml saline solution) and injection is performed with 1.5 ml luer lock syringe with a 27G needle injections above the periosteum (Fig. 2.16).

Are deep fat compartments enhancement needed? Typically, at this age, enhancing deep compartments is necessary in the malar area and inferior eyelid deep compartment the Suborbicularis Oculi Fat (SOOF). SOOF is a distinct layer of fat located just beneath the orbicularis oculi muscle in the lower eyelid. SOOF plays a crucial role in facial aesthetics, especially in the context of aging, by providing support and cushioning to the lower eyelid.

The treatment of the deep fat compartment of malar area involves the injection of 2–3 ml of microfat (no passages of emulsification), with a 3 ml luer lock syringe and 18G microcannula; the injection is performed linear and retrograde beneath the SMAS, deeper than for superficial compartments.

The treatment of the SOOF involves injecting 1–1.5 ml of tissue at 500 microns (5–6 passages of emulsification) beneath the orbicularis muscle using a 21G microcannula. Prior to injection, administering a small amount of local anesthesia with adrenaline (1:100,000) is advisable. The entry point is the lateral inferior border of the orbital frame; the cannula is inserted perpendicularly, touching the bone, then

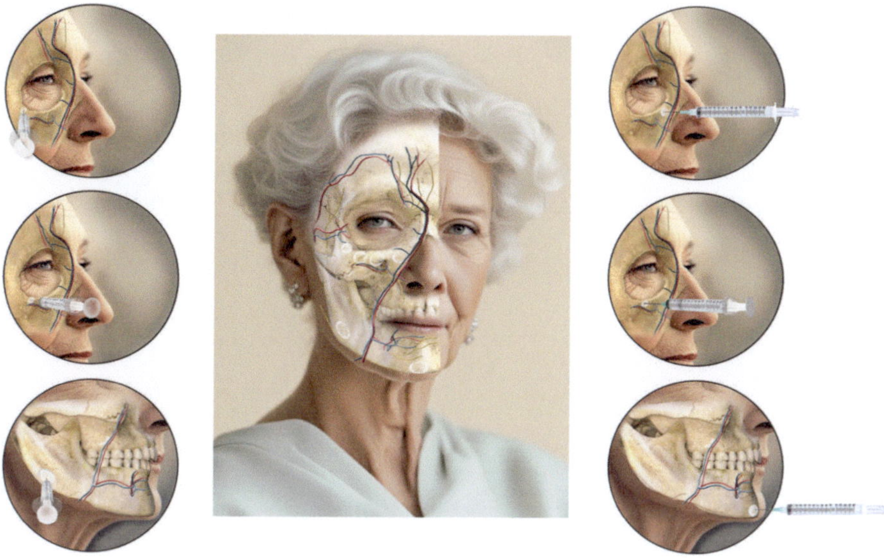

Fig. 2.16 CaHA deep injections above the periosteum

rotated 90 degrees, and slid under the orbicularis muscle. The injection is linear and retrograde (Fig. 2.17).

Does she require superficial fat compartment enhancement? (Fig. 2.18) At her age, these compartments are mostly impacted by aging.

In the upper third, treatment of superficial compartments includes the tail of the eyebrow and the temporal area. Here, I plan to inject 2–4 ml of tissue at 600 or 500 microns (2–3 passages of fluidification), depending on skin thickness, using a 20G cannula. The injections in all areas are performed linearly and retrograde, following an initial administration of a small amount of local anesthesia with adrenaline (1:100,000). The injection in the eyebrow is subcutaneous, while in the temporal area, it is above the major fascia of the temporal muscle. In the temporal area, the entry point is made caudal to the temporal crest; the cannula slides over the fascia, starting the injection from the caudal part of the eyebrow and gradually moving downwards to the temporal area, up to the orbital frame and zygomatic arch.

In the middle third, the treatment targets the malar fat pad and the nasolabial fold.

Enhancing the superficial fat pad in the malar area is key to correcting mid-face deflation, which causes tissue descent and nasolabial fold prominence.

An anesthetic block of the infraorbital nerve is often considered. Typically, I inject 4–6 ml of tissue at 600 microns (2–3 passages of fluidification) using a 20G cannula. The entry points are both in the inferior and superior lateral malar area; the injection is in the superficial subcutaneous layer and performed using a linear retrograde technique; the linear injections performed from the two-entry points form "net" pattern.

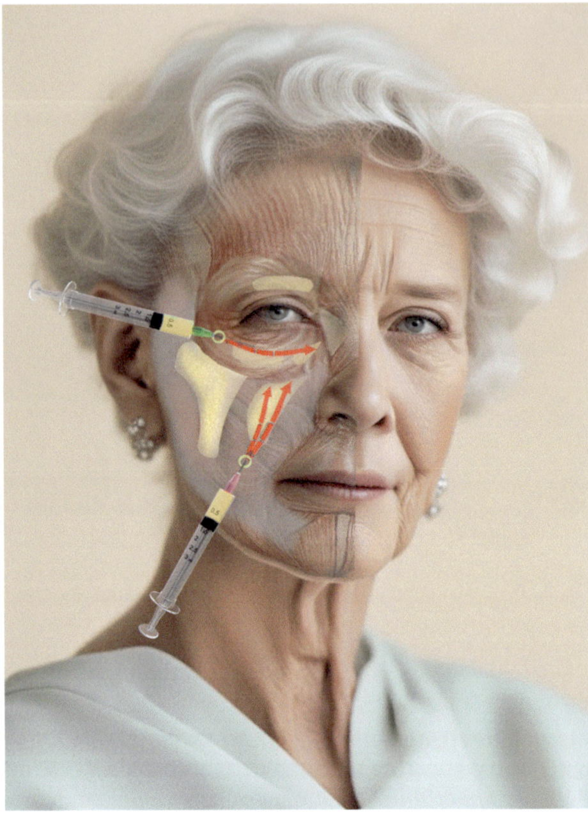

Fig. 2.17 Tissue injection of the deep fat compartments

In the lower third of the face, treatments are focused on the perioral region and jawline. I typically address the "Marionette" area by injecting 2 to 2.5 milliliters of tissue at 600 microns (2–3 passages of fluidification) using a 20G cannula, with my preferred entry point being the center of the chin. For the jawline and preauricular area, I favor the mandibular angle as the entry point. Here, I administer 3 to 4 milliliters of tissue at 600 microns using a 20G cannula. In the preauricular area, the usual approach involves injecting 2 to 2.5 milliliters of tissue at 500 to 600 microns, utilizing either a 20G or 21G cannula.

After 55 to Over 65 Years (Fig. 2.19): Aging signs become more dynamic and exaggerated:

Forehead and glabellar lines deepen and are evident even at rest.

Further drooping of the upper eyelid, increased tear trough severity, and nasal drooping.

Noticeable descent of midface structures and potential hollowing due to tooth loss.

Increased prominence of nasolabial folds and perioral lines, with a loss of jawline definition and development of jowls.

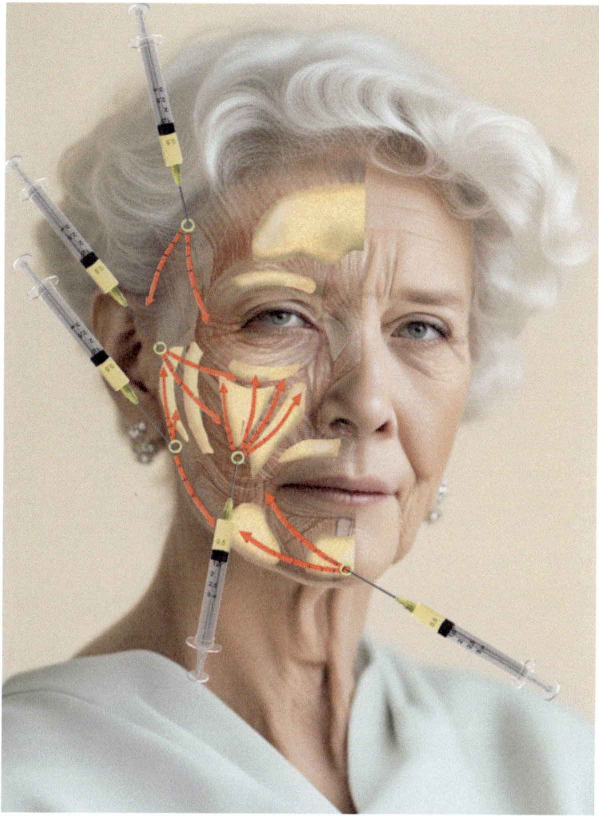

Fig. 2.18 Tissue injection of the superficial fat compartments

The treatment approach at this stage must be comprehensive, often combining multiple modalities to address the extensive range of aging signs.

Does she require bone support? At her age, a bone support is always necessary, in particular, in the upper part of the malar and zygomatic area, in the jawline, and in the chin.

In the upper part of the malar and zygomatic area, my plan includes injecting multiple micro boluses of CaHA (1.5 ml of CaHA diluted in 0.5 ml saline solution) into the medial malar area, gradually decreasing the volume towards the zygomatic area; The injection technique involves a 1.5 ml luer lock syringe with a 27G needle, performing deep injections above the periosteum.

In in the jawline mandibular angle, the injection involves two to five micro boluses of CaHA (1.5 ml of CaHA diluted in 0.5 ml saline solution) from the mandibular angle to the medial part of the mandibular and injection is performed with 1.5 ml luer lock syringe with a 27G needle injection above the periosteum. In this area is mandatory to mark on the skin both the branch of the facial artery and vein and the mental nerve in order to inject filler away from these anatomical structures:

Fig. 2.19 55 - to 65-year-old patient

- The Mandibular Branch of the Facial Artery: The facial artery is one of the major arteries supplying blood to the face. Its mandibular branch, specifically, has a distinct course and is important in both clinical and surgical contexts. The facial artery is a branch of the external carotid artery. After originating, it takes a tortuous course upwards and forwards to the face.
  The mandibular branch of the facial artery typically runs superficial to the submandibular gland and then hooks around the lower border of the mandible. This course is usually just anterior to the masseter muscle and at the anteroinferior angle of the masseter. The mandibular branch is deep to the platysma and superficial to the buccinator muscle. In its course, it may be closely associated with the facial vein and several cranial nerve branches, particularly the marginal mandibular branch of the facial nerve. Its pulsation can often be felt and is used as a landmark in facial procedures.
- The Mental Nerve: It is a sensory nerve branch of the mandibular division of the trigeminal nerve (cranial nerve V3). It is essential for sensation in the lower face. The nerve travels through the mandibular canal within the lower jaw. It exits the mandible through the mental foramen, which is typically located below the second premolar tooth. Upon exiting the mental foramen, it divides into several terminal branches. These branches provide sensory innervation to the lower lip, chin, and buccal gingiva (gums) of the anterior teeth.

In the chin, the injection involves one to three micro boluses of CaHA (1.5 ml of CaHA diluted in 0.5 ml saline solution), and injection is performed with 1.5 ml luer lock syringe with a 27G needle injections above the periosteum (Fig. 2.20). The chin bone reabsorption negatively affects the neck angle aspect, and its volume enhancement leads to an important improvement in the lower third of the face.

Are deep fat compartments enhancement needed? (Fig. 2.21) At this age, enhancing deep compartments is necessary in the malar area and both inferior (SOOF) and upper (ROOF) eyelid. The retro-orbicularis oculi fat (ROOF) in the upper eyelid refers to a specific anatomical structure in the eyelid and orbital area. The ROOF is situated just above the orbicularis oculi muscle in the upper eyelid; it serves as a cushioning layer for the orbital contents and plays a role in the aesthetic contour of the upper eyelid. It helps in maintaining the smooth appearance of the eyelid and the upper periocular region.

The treatment of the deep fat compartment of malar area involves the injection of 3–4 ml of microfat (no passages of emulsification), with a 3 ml luer lock syringe and 18G microcannula; the injection is performed linear and retrograde beneath the SMAS, deeper than for superficial compartments.

The treatment of the SOOF involves injecting 1–1.5 ml of tissue at 500 microns (5–6 passages of emulsification) beneath the orbicularis muscle using a 21G microcannula. Prior to injection, administering a small amount of local anesthesia with adrenaline (1:100,000) is advisable. The entry point is the lateral inferior border of the orbital frame; the cannula is inserted perpendicularly, touching the bone, then rotated 90 degrees, and slid under the orbicularis muscle. The injection is linear and retrograde.

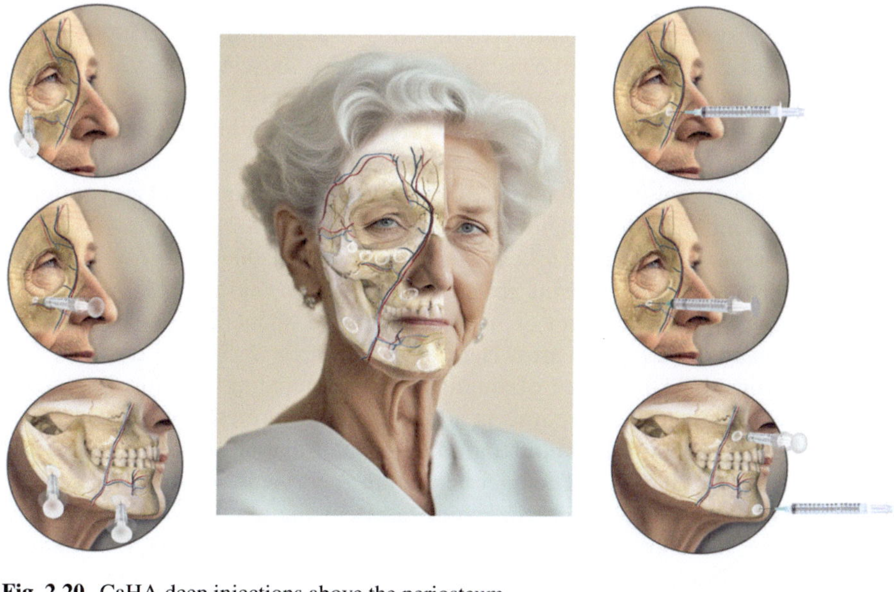

Fig. 2.20 CaHA deep injections above the periosteum

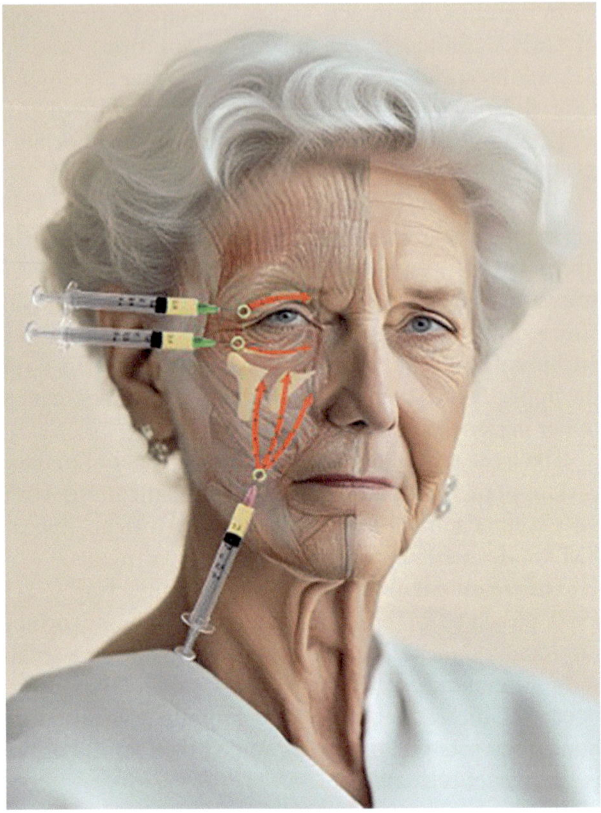

Fig. 2.21 Tissue injection of the deep fat compartments

The treatment of the ROOF involves injecting 1–2 ml of tissue at 500 microns (5–6 passages of emulsification) beneath the orbicularis muscle using a 21G microcannula. Prior to injection, administering a small amount of local anesthesia with adrenaline (1:100,000) is advisable. The entry point is the lateral surface of the temporal area in front of the hairline; the cannula is inserted perpendicularly, touching the major temporal fascia, then slides towards the orbit frame, and moves under the orbicularis muscle. The injection is linear and retrograde.

Does she require superficial fat compartment enhancement? (Fig. 2.22).

At her age, these compartments are mostly impacted by aging.

In the upper third, treatment of superficial compartments includes the tail of the eyebrow and the temporal area. Here, I plan to inject 2–5 ml of tissue at 600 or 500 microns (2–3 passages of fluidification), depending on skin thickness, using a 20G cannula. The injections in all areas are performed linearly and retrograde, following an initial administration of a small amount of local anesthesia with adrenaline (1:100,000). The injection in the eyebrow is subcutaneous, while in the temporal area, it is above the major fascia of the temporal muscle. In the temporal area, the entry point is made caudal to the temporal crest; the cannula slides over the fascia, starting the injection from the caudal part of the eyebrow

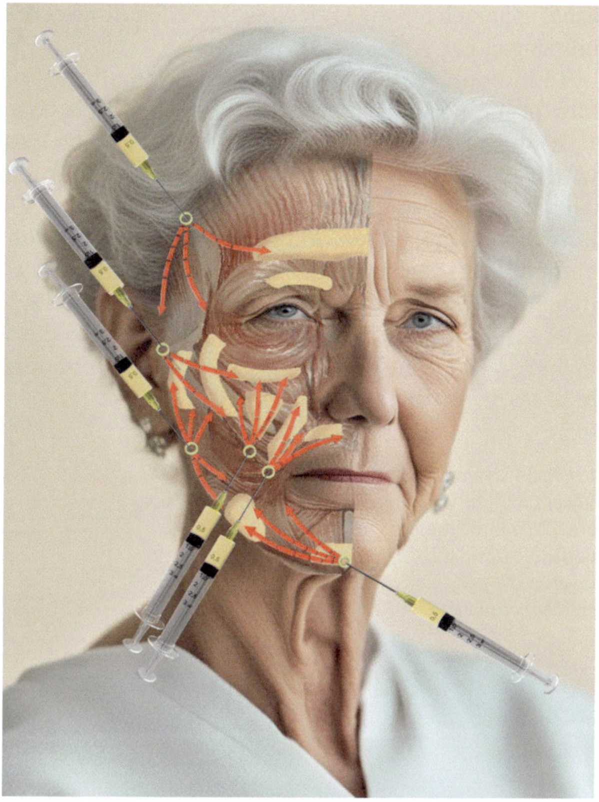

Fig. 2.22 Tissue injection of the superficial fat compartments

and gradually moving downwards to the temporal area, up to the orbital frame and zygomatic arch.

In the middle third, the treatment targets the malar fat pad and the nasolabial fold.

Enhancing the superficial fat pad in the malar area is key to correcting mid-face deflation, which causes tissue descent and nasolabial fold prominence.

An anesthetic block of the infraorbital nerve is often considered. Typically, I inject 6–8 ml of tissue at 600 microns (2–3 passages of fluidification) using a 20G cannula. The entry points are both in the inferior and superior lateral malar area; the injection is in the superficial subcutaneous layer and performed using a linear retrograde technique; the linear injections performed from the two entry points form a "net" pattern.

In the lower third of the face, treatments are focused on the perioral region and jawline. I typically address the "Marionette" area by injecting 2 to 2.5 milliliters of tissue at 600 microns (2–3 passages of fluidification) using a 20G cannula, with my preferred entry point being the center of the chin. For the jawline and preauricular area, I favor the mandibular angle as the entry point. Here, I administer 4 to 5 milliliters of tissue at 600 microns using a 20G cannula. In the preauricular area, the usual approach involves injecting 2 to 2.5 milliliters of tissue at 500 to 600 microns, utilizing either a 20G or 21G cannula.

In individuals between the ages of 55 and 65, the emergence of fine wrinkles in the perioral area, often described as a "barcode" pattern, is a typical sign of aging. The treatment of this area involves the injection of nanofat, which is processed through 30 stages of fluidification. This is administered using a 1 ml luer lock syringe and a 27G needle. The injection is performed into the dermal layer or the very superficial subdermal plane (Fig. 2.23). Targeting the dermal layer is effective for treating fine lines and wrinkles, as it enables the rejuvenative elements of the nanofat to directly interact with the skin cells. In certain instances, injecting into the very superficial subdermal layer can prove advantageous, particularly for addressing slightly deeper lines or areas requiring more volume.

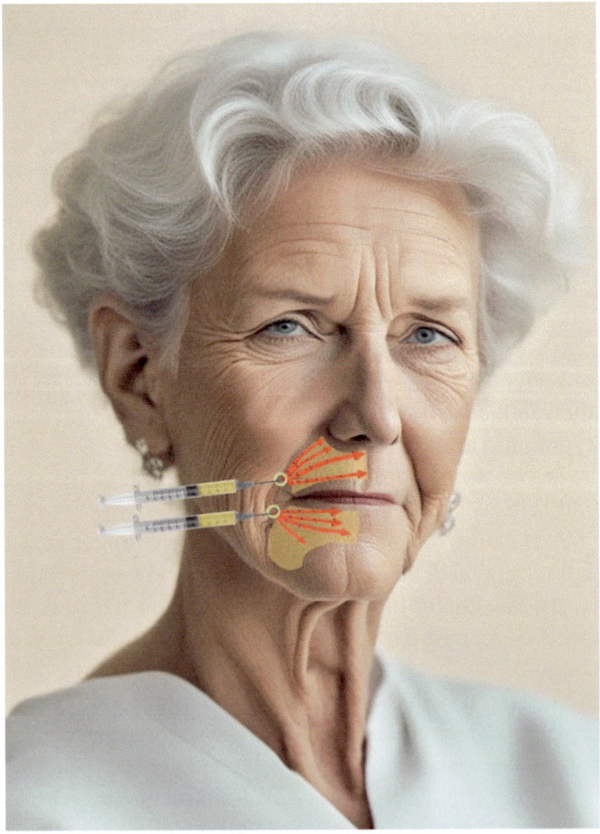

Fig. 2.23 Nanofat injection in the perioral area: The injection is performed into the dermal layer or the very superficial subdermal plane

2.17 Caution in Facial Injection

Dermal fillers have been injected with increasing frequency over the past three decades for soft tissue augmentation by volume expansion in the management of the aging face. Over the past several years, the number of procedures involving soft tissue fillers has increased from 1.6 million per year in 2011 to more than 2.4 million in 2015. This widespread use has led to a rise in reports of associated complications [74].

Additionally, with the global increase in fat grafting procedures, there has been a rise in the number of reported cases of complications, included vascular complications. In this review, vascular complications associated with facial fat grafting are examined [75].

For both procedures (fillers and fat grafts), the complication rate is very low. However, the rarest but most feared complication of injectable procedures in the face is vascular complications. Although ischemic events are exceptionally rare, they represent a small percentage of complications in individual clinical practices. Typical clinical findings include skin blanching, livedo reticularis, slow capillary refill, and dusky blue-red discoloration, followed a few days later by blister formation and ultimately tissue sloughing. Due to arterial compromise, the pathology of vascular complications involves tissue anoxia and possible progression to necrosis. The mainstays of treatment (apart from avoidance through meticulous technique) include prompt recognition, immediate treatment with hyaluronidase (for HA fillers), topical nitro paste under occlusion, oral acetylsalicylic acid (aspirin), warm compresses, and vigorous massage. Secondary lines of treatment may involve intra-arterial hyaluronidase (for HA fillers), hyperbaric oxygen therapy, and ancillary vasodilating agents such as prostaglandin E1. Vascular complications also include retrograde vascular embolization of the ophthalmic artery that can cause blindness [67].

In this systematic review of the published literature to assess the association of blindness with filler and fat injection: between the January 2000 and September 2018 with a total of 190 cases of filler-induced blindness or other ocular disturbances, 90 cases were attributed to autologous fart injections **(47%)** and 100 cases from fillers **(53%)** (HA28%, CaHA4%, others21%).

The percentage of reported filler-associated blindness according to the injection site is as follows [76]:

- Glabella complex 34%
- Nose 24%
- Nasolabial fold 12%
- Forehead 12%
- Temple 7%
- Cheek 6%
- Periorbital area 3%
- Chin 1%
- Eyebrow 1%

In this retrospective study conducted between January 1, 2008, and August 31, 2014, it is observed that HA filler is more likely to block the central retinal artery (approximately 160 μm in size) or its smaller branches and fat more likely the ophthalmic artery (2 mm in diameter) due to the particle size and the amount of filler [77].

2.18 Intravascular Complications in Facial Adipose Tissue Grafting: Focused Insights and Management

This study is a systematic search of the literature was performed interrogating PubMed search engines until July 2021 [3]. In total, 462 papers were identified by PubMed search. A total of 298 adverse events were identified: 40 (13.4%) intravascular injections, 13 (4.3%) asymmetry, 57 (19.1%) irregularities, 22 (7.4%) graft hypertrophy, 21 (7%) fat necrosis, 73 (24.5%) prolonged edema, 1 (0.3%) infection, 6 (2%) prolonged erythema, 15 (5%) telangiectasia, and 50 (16.8%) cases of acne activation.

The highest rate of vascular complications occurred in cases of multisite injections (16–18.4%).

Risk-based facial map:

- Glabella treatment (16–18.4%)
- Forehead (10–11.5%)
- Temporal area (8–9.2%)
- Periocular region, nose, and nasolabial folds carried a medium-low risk (5–5.7%, 4–4.6%, and 4–4.6% risk, respectively).

The safest facial zone to inject was the cheek [73].

This other study is a review of the published literature before May 2020 to assess the association between vascular complications and fat grafting. In the present study, we identified a total of 111 cases of fat-induced severe vascular complications in the literature. A total of 68 patients with ophthalmic artery embolism after fat grafting were identified in this study.

Risk-based facial map [75].

- Glabella 26%
- Forehead 18.7%
- Temple 19.5%
- Nose 7%
- Nasolabial fold 6%
- Periocular 5.7%
- Cheek 1.6%

2.19 Anatomical Consideration

As previously described, the rare but most serious complications of filler or adipose tissue injections are arterial embolization, which can lead to skin necrosis and blindness. Skin necrosis occurs due to embolization of a terminal branch of the facial arteries, while blindness results from retrograde embolization into the ophthalmic arteries.

Two main arterial circulations are involved in these vascular complications: arteries arising from the external carotid artery (Fig. 2.24) and those from the internal carotid artery (Fig. 2.25). Understanding the connections between these two systems is crucial to prevent retrograde embolization, the primary cause of the most feared complication, blindness.

In particular, both the facial artery and the superficial temporal artery arise from the external carotid artery, while the ophthalmic artery arises from the internal carotid artery.

2.19.1 Skin Necrosis

The complication that leads to skin necrosis (Fig. 2.26 primarily results from the embolization of the terminal branches of the facial artery (Fig. 2.26), as follows:

- Superior and Inferior Labial Arteries: Embolization of these arteries can cause skin necrosis of the upper and lower lips.

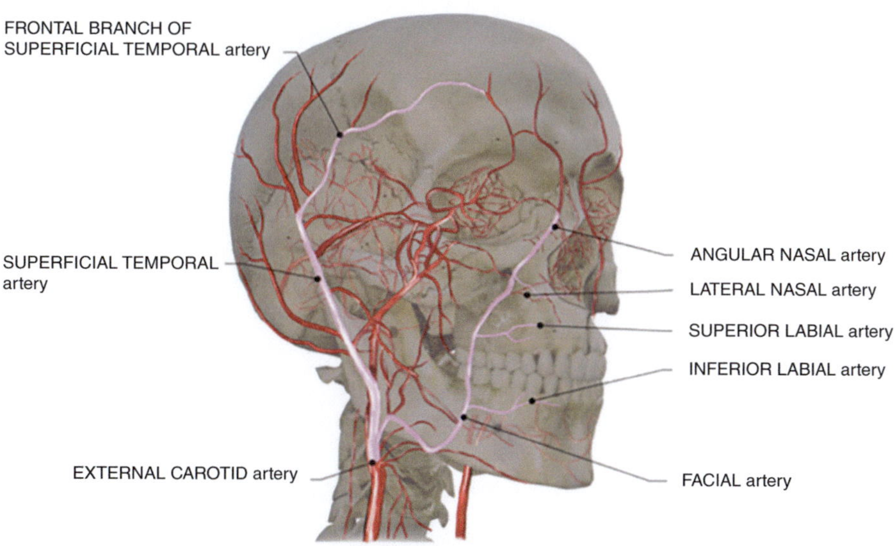

Fig. 2.24 Pattern of facial external carotid circulation

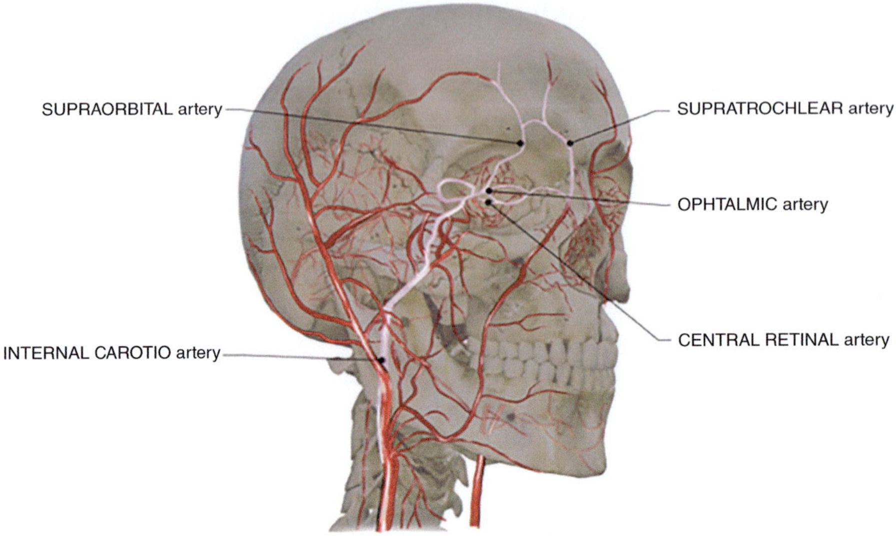

Fig. 2.25 Pattern of facial internal carotid circulation

Fig. 2.26 Dangerous area for skin necrosis

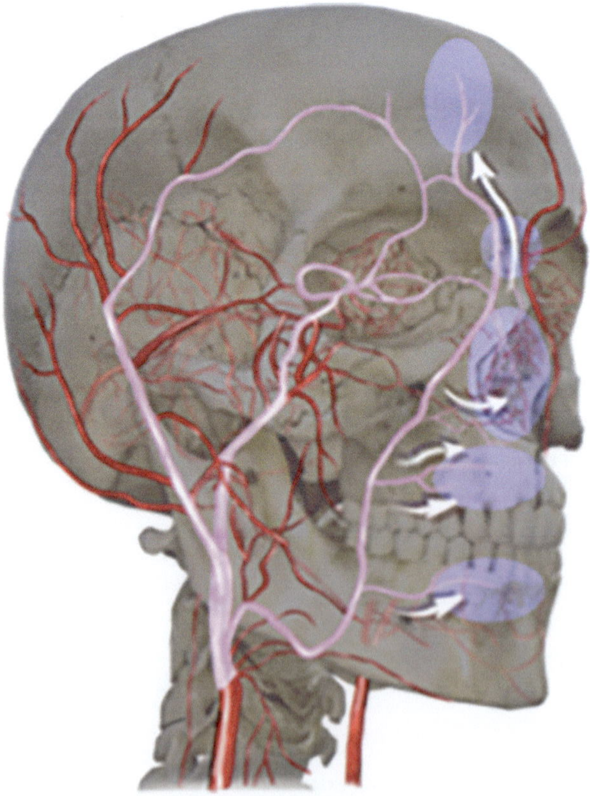

- Lateral Nasal Artery: Embolization can cause necrosis of the alar and dorsum of the nose.

Additionally, embolization of the supratrochlear artery, a branch of the ophthalmic artery arising from the internal carotid artery, can cause necrosis of the skin in the glabella area and the medial portion of the forehead.

2.19.2 Blindness

The normal blood flow direction in the external and internal carotid arteries and their branches is anterograde, meaning that the blood flows from the heart outward through the carotid arteries to their peripheral branches. During the injection of a filler or adipose tissue, blindness can result from the retrograde embolization of the injected material. This occurs when the embolus, pushed by the force of the injection, travels backward against the normal anterograde blood flow. The embolus moves from the branches of the external carotid artery into the ophthalmic artery, which is a branch of the internal carotid artery. This retrograde movement can lead to occlusion of the central retinal artery or its branches, resulting in sudden and potentially irreversible vision loss.

Taking these considerations into account, we can identify two main gateways for retrograde embolization (Fig. 2.27): the medial superior border of the orbit (supratrochlear and supraorbital arteries) and the superior portion of the nasal bone (angular and ophthalmic arteries). In the next paragraph, we suggest applying pressure

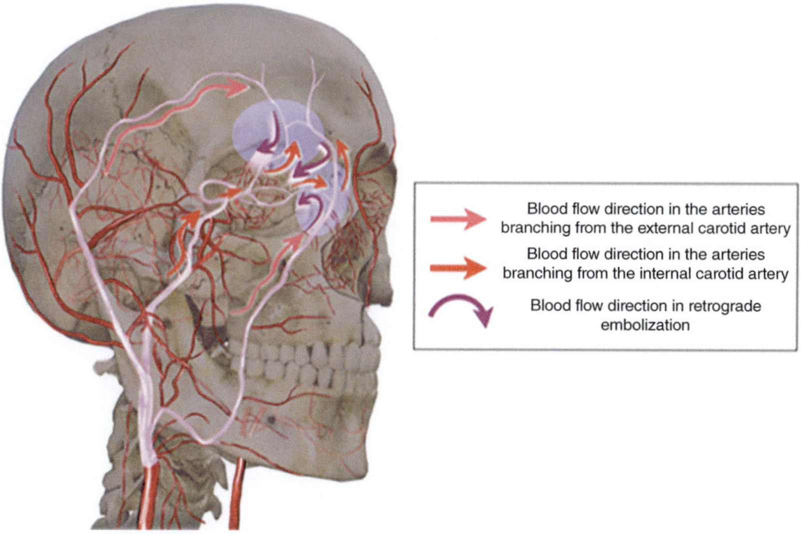

Fig. 2.27 Gateways for retrograde embolization

with the nondominant hand to these areas during injections in potentially dangerous zones.

2.19.3 *Caution Zones*

Identifying caution zones in the face for injections, such as fillers or fat, is crucial for avoiding complications like embolization of the facial artery branches. These key zones are typically where arteries are either superficial or in close proximity to common injection sites (Fig. 2.28). Here are the primary areas of concern:

- **Glabella and Forehead**: Injections here are risky due to the supratrochlear and supraorbital arteries. These areas have limited collateral circulation, increasing the risk of necrosis; moreover, these arteries are branches of ophthalmic artery; then if retrograde embolization occurs, it causes blindness.
- **Nasal Region**: The nose has a rich blood supply from branches of both the external and internal carotid arteries, such as the lateral nasal artery (from the facial artery), the dorsal nasal artery (from the ophthalmic artery), and the angular artery (from facial artery). Accidental injection into these arteries can lead to serious complications as skin necrosis and blindness due to retrograde embolization in the ophthalmic artery.
- **Nasolabial Fold**: This region is supplied by the facial artery. Care must be taken to avoid deep injections which can affect the artery, causing necrosis and retrograde embolization.

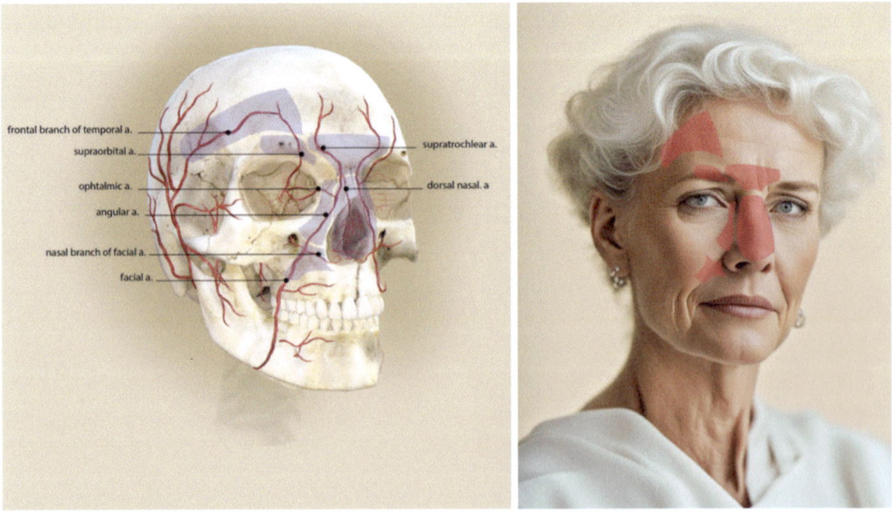

Fig. 2.28 Caution areas

- **Periocular and Temporal Areas:** Where the superficial temporal artery and the infraorbital artery can be affected. Particular attention must be paid during injections in the temporal area; embolization of the superficial temporal artery or its frontal branch can cause retrograde embolization through the supraorbital or supratrochlear arteries, which are branches of the ophthalmic artery.

It is also crucial for practitioners to be aware of early signs of vascular compromise, which include blanching of the skin, severe pain, and changes in vision, to initiate prompt and appropriate management [67–71].

2.20 Prevention Vascular Complications

The main causes of vascular embolism are vascular rupture, high injection pressure, and excessive single-point injection dose.

2.20.1 Prevention of Vascular Rupture

1. Know the location and depth of facial vessels and the common variations. Injectors should understand the appropriate depth and plane of injection at different sites.
2. Use local anesthetics and epinephrine to contract blood vessels.
3. Use larger, blunter cannula. The volume placed with each pass of the cannula should be less than 0.1 mL.
4. Use extreme caution when injecting a patient who has undergone a previous surgical procedure in the area.

2.20.2 Prevention of High Injection Pressure

1. Withdraw the cannula during injection and inject slowly with minimal pressure.
2. Operate gently during injection.

2.20.3 Prevention of Excessive Single-Point Injection Dose

1. Limit the syringe size 1 to 3 mL.
2. Inject small amounts during cannula extraction; no more than 0.1 mL of tissue should be injected with each increment.

3. Move the tip of the cannula while injecting (fan-shaped retrograde linear technique) to avoid depositing a large amount in a single location.
4. Do not force the plunger in case of the cannula clog but change the cannula.

2.20.4 Additional Maneuver

We can identify two main gateways for retrograde embolization: the medial superior border of the orbit (supratrochlear and supraorbital arteries) and the superior portion of the nasal bone (angular and ophthalmic arteries).

We suggest applying pressure with the nondominant hand to these areas during injections in potentially dangerous zones (Figs. 2.29, 2.30, 2.31, and 2.32).

In case of treating dangerous areas, the vascular mapping with Doppler ultrasound could be useful [74, 75].

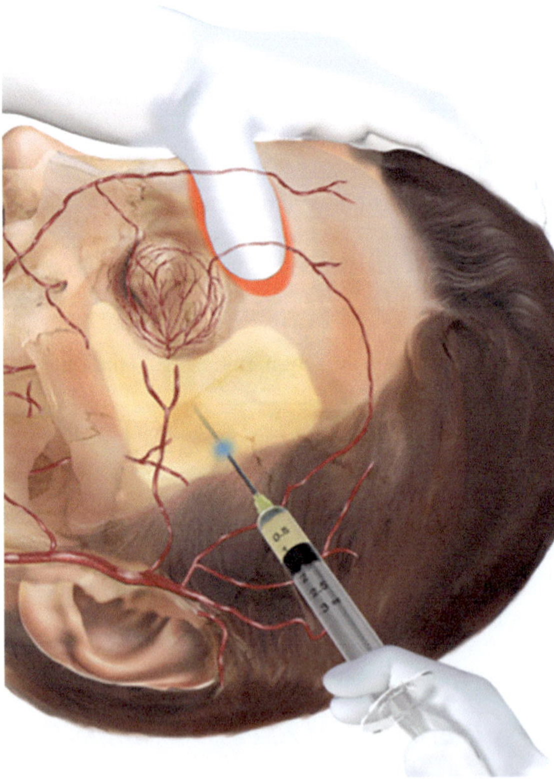

Fig. 2.29 Left temporal area treatment: during the tissue injection, the thumb of the no dominant hand makes a pressure on the supratrochlear and supraorbital arteries

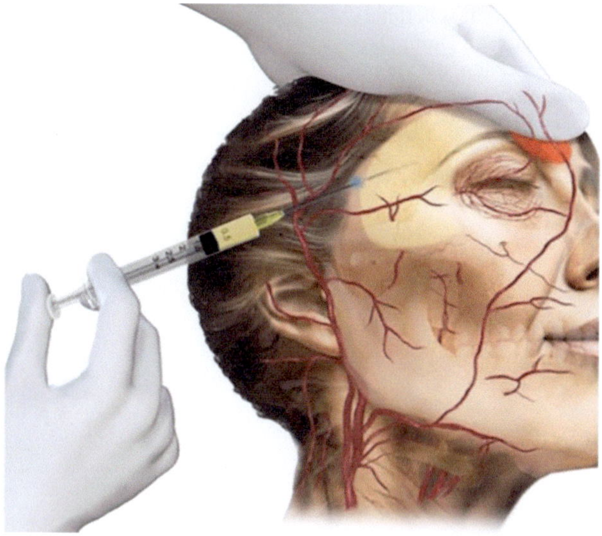

Fig. 2.30 Right temporal area treatment: During the tissue injection, the thumb of the no dominant hand makes a pressure on the supratrochlear and supraorbital arteries

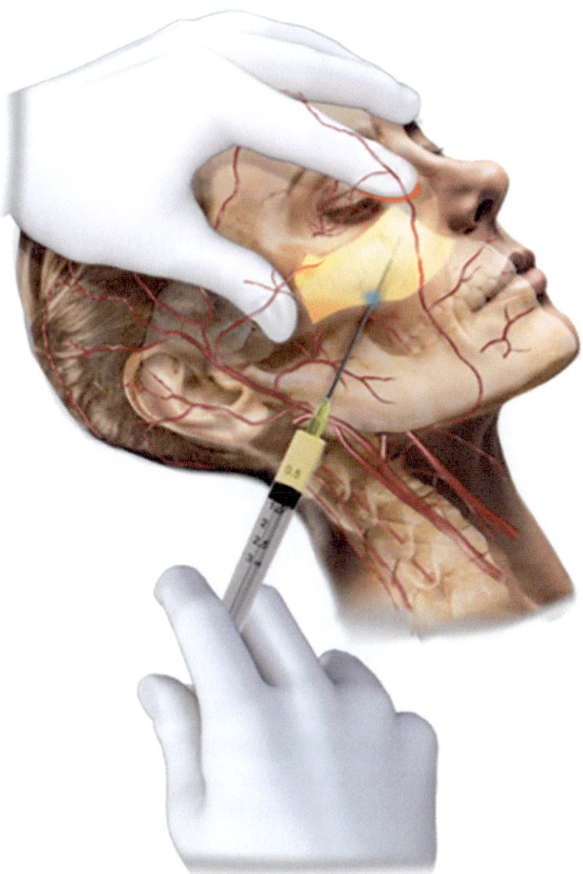

Fig. 2.31 Right suborbital area treatment: During the tissue injection, the thumb of the no dominant hand makes a pressure on the angular artery

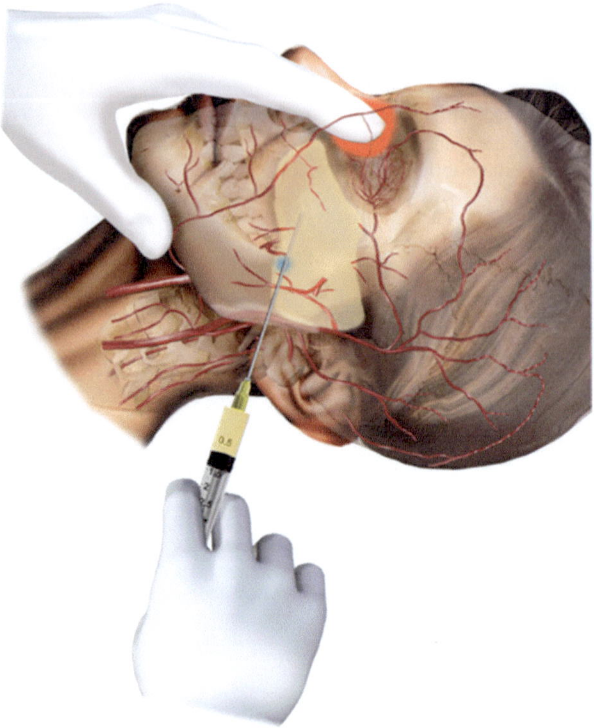

Fig. 2.32 Left suborbital area treatment: During the tissue injection, the thumb of the no dominant hand makes a pressure on the angular artery

2.20.5 A Comprehensive Evaluation of the Safety of Guided SEFFI Technique in Facial Autologous Regenerative Therapy

Facial rejuvenation through regenerative therapy, particularly using the guided Superficial Enhanced Fluid Fat Injection (SEFFI) technique, has emerged as a significant advancement in aesthetic medicine. This technique involves the autologous homologous transplantation of adipose tissue, ensuring a high degree of biocompatibility and safety. The guided SEFFI procedure, which employs a specialized guide for adipose tissue harvesting, has been designed to be accessible even to practitioners without extensive surgical expertise, ensuring a safe, simple, standardized, and effective approach. In a landmark, Italian multicenter retrospective case report investigated the safety of this procedure: this study collected data from 2365 procedures performed in Italian centers between 2019 and 2021. Guided SEFFI was performed alone or combined with cosmetic treatments, including the use of hyaluronic acid filler, suspension threads, synthetic calcium hydroxylapatite, botulin toxin, and microneedling. Results: guided SEFFI was used alone in more than 60% of the patients and in all facial areas. In about one-tenth of the patients, guided SEFFI was combined with a botulin toxin treatment or hyaluronic acid filling. Other procedures were used more rarely. Ecchymosis in the donor or injection sites was the most frequent adverse event but was only observed in 14.2% and 38.6% of the patients, respectively. In conclusion, the guided SEFFI technique is standardized and minimally invasive, leading to very few complications. (Table 2.1) It constitutes a promising antiaging medical treatment that combines effectiveness, safety, and simplicity [72].

A comprehensive review of complications associated with facial fat grafting, encompassing 5479 patients, indicated an average complication rate varying significantly across studies, with some adverse events noted [73].

These complications were categorized into severe, moderate, and minor, based on their impact and required management approach. However, the results from the extensive guided SEFFI study, which included 2365 patients, presented a reassuring safety profile with no severe or moderate side effects reported. The minor side effects, predominantly ecchymosis at the donor (Table 2.2) and injection sites (Table 2.3), were relatively infrequent and manageable.

Our comprehensive and meticulously conducted study confirms the safety of the guided SEFFI procedure. Among the 2365 patients evaluated, no moderate or severe side effects were observed. Only minor adverse events were noted, such as ecchymosis at both the donor site (14.2%) and injection site (38.6%). Other complications, including hematoma or persistent skin irregularities lasting over 20 days, were rare occurring in just 1.3% and 0.2% of cases, respectively. These findings strongly support the safety profile of the procedure, with ecchymosis being the only commonly reported minor issue. The low incidence of additional complications further underscores its safety. The minimal complication rate at the donor site is likely due to the relatively small volume of tissue harvested, especially when compared to traditional liposuction or harvesting methods reported in earlier studies [73, 78].

In conclusion, the presented study proves that guided SEFFI alone or in combination with other facial rejuvenation techniques performed in medical facilities is a minimally invasive procedure with a very low rate of complications.

Table 2.1 A total of 2365 procedures was investigated: 63% were single procedure and 37% were combined procedures: this table describes the specific combined procedures

Guided SEFFI procedures		Number of procedures	Percentage of patients
Single procedure		1484	62.7
Combined procedures	Botulin toxin	295	12.5
	Hyaluronic acids	233	9.9
	Threads	122	5.2
	Microneedling	85	3.6
	Calcium hydroxylapatite	65	2.7
	Carboxytherapy	51	2.6
	Endolift	24	1.0
	Other	0	0
	Total	875	37.0

Table 2.2 Complications rate observed in the donor site

Site of complication	Type of complication	Number of events	Percentage of patients
Donor site	Ecchymosis	336	14.2
	Hematoma	31	1.3
	Prolonged erythema (>48 h)	0	0
	Skin necrosis	0	0
	Prolonged edema (>20 days)	0	0
	Infection	0	0
	Fat necrosis	0	0
	Telangiectasias	0	0
	Skin irregularities (>20 days)	5	0.2
	Embolism	0	0
	Other	0	0

Table 2.3 Complications rate observed in the injection site

Site of complication	Type of complication	Number of events	Percentage of patients
Injection site	Ecchymosis	914	38.6
	Hematoma	18	0.8
	Prolonged erythema (>48 h)	0	0
	Prolonged erythema (>48 h)	0	0
	Skin necrosis	0	0
	Prolonged edema (>5 days)	14	0.6
	Infection	1	<0.1
	Fat necrosis	0	0
	Telangiectasias	0	0
	Activation of acne	0	0
	Visibility	2	0.1
	Skin irregularities (>48 h)	1	<0.1
	Blindness	0	0
	Asymmetry	3	0.1
	Other	0	0

Table 2.4 Distribution of donor sites: numbers and percentages

Donor site	Average harvested volume (mL)	Number of patients (%)
abdomen	27.1	1231 (52.1)
trochanteric	27.1	761 (32.4)
hip	22.6	296 (12.6)
knee	6.8	67 (2.8)
others	2.1	3 (0.1)

The study also highlighted the procedural specifics, noting the abdomen and trochanteric site as the primary harvesting areas, with an average tissue collection of 27.1 mL, underscoring the minimally invasive nature of the technique (Table 2.4).

The treatments were predominantly focused on the malar-zygomatic area, with an average of 4.4 mL tissue injected per side. Guided SEFFI was often performed as a singular treatment, but it also proved effective when combined with other techniques, such as botulinum toxin and hyaluronic acid injections.

In conclusion, this comprehensive analysis of the guided SEFFI technique in facial Autologous Regenerative Therapy demonstrates its potential as a safe, effective, and minimally invasive approach to facial rejuvenation. Its low complication rate, coupled with the simplicity of the procedure, marks it as a promising option in the field of medical antiaging treatments.

2.21 Conclusion

Regenerative facial rejuvenation demands a clear understanding of the aging process and its variations across different age groups. A thorough, inside-out evaluation of each patient is essential to identify which anatomical compartments require restoration. Autologous Regenerative Therapy in Aesthetics (ARTA), performed using the guided SEFFI technique, has demonstrated effectiveness, safety, simplicity, and standardization. This procedure can be carried out in medical facilities without the need for surgical or liposuction expertise.

To achieve optimal outcomes, ARTA should be integrated with other medical or surgical treatments. The central message of this book is that the first step toward harmonious facial rejuvenation is the regeneration of tissues in the most natural and safe way—using the patient's own tissue. This patient-centered approach ensures that rejuvenation strategies are both effective and tailored to the individual's specific stage of facial aging.

References

1. Cotofana S, Fratila AA, Schenck TL, Redka-Swoboda W, Zilinsky I, Pavicic T. The anatomy of the aging face: a review. Facial Plast Surg. 2016;32(3):253–60. https://doi.org/10.1055/s-0036-1582234. Epub 2016 Jun 1. PMID: 27248022.
2. Wan D, Amirlak B, Giessler P, Rasko Y, Rohrich RJ, Yuan C, Lysikowski J, Delgado I, Davis K. The differing adipocyte morphologies of deep versus superficial midfacial fat compartments: a cadaveric study. Plast Reconstr Surg. 2014;133(5):615e–22e. https://doi.org/10.1097/PRS.0000000000000100. PMID: 24445875.
3. Sykes JM, Cotofana S, Trevidic P, et al. Upper face: clinical anatomy and regional approaches with injectable fillers. Plast Reconstr Surg. 2015;136(5, Suppl):204S–18S.
4. Shaw RB Jr, Katzel EB, Koltz PF, Kahn DM, Girotto JA, Langstein HN. Aging of the mandible and its aesthetic implications. Plast Reconstr Surg. 2010;125(1):332–42.
5. Toledo Avelar LE, Cardoso MA, Santos Bordoni L, de Miranda Avelar L, de Miranda Avelar JV. Aging and sexual differences of the human skull. Plast Reconstr Surg Glob Open. 2017;5(4):e1297.
6. Shaw RB Jr, Katzel EB, Koltz PF, et al. Aging of the facial skeleton: aesthetic implications and rejuvenation strategies. Plast Reconstr Surg. 2011;127(1):374–83.

7. Pessa JE. An algorithm of facial aging: verification of Lambros's theory by three-dimensional stereolithography, with reference to the pathogenesis of midfacial aging, scleral show, and the lateral suborbital trough deformity. Plast Reconstr Surg 2000;106(2):479–488, discussion 489.
8. Richard MJ, Morris C, Deen BF, Gray L, Woodward JA. Analysis of the anatomic changes of the aging facial skeleton using computer-assisted tomography. Ophthalmic Plast Reconstr Surg. 2009;25(5):382–6.
9. Swift A, Liew S, Weinkle S, Garcia JK, Silberberg MB. The facial aging process from the "inside out". Aesthet Surg J. 2021;41(10):1107–19. https://doi.org/10.1093/asj/sjaa339. PMID: 33325497; PMCID: PMC8438644.
10. Lambros VMD. Observations on Periorbital and Midface Aging. Plast Reconstr Surg. 2007;120(5):1367–76. https://doi.org/10.1097/01.prs.0000279348.09156.c3.
11. Rohrich RJ, Avashia YJ, Savetsky IL. Prediction of facial aging using the facial fat compartments. Plast Reconstr Surg. 2021;147(1S-2):38S–42S. https://doi.org/10.1097/PRS.0000000000007624. PMID: 33347073.
12. Rohrich RJ, Arbique GM, Wong C, et al. The anatomy of suborbicularis fat: implications for periorbital rejuvenation. Plast Reconstr Surg. 2009;124:946–51.
13. Rohrich RJ, Pessa JE. The anatomy and clinical implications of perioral submuscular fat. Plast Reconstr Surg. 2009;124:266–71.
14. Gierloff M, Stöhring C, Buder T, et al. Aging changes of the midfacial fat compartments: a computed tomographic study. Plast Reconstr Surg. 2012;129:263–73.
15. Owsley JQ, Roberts CL. Some anatomical observations on midface aging and long-term results of surgical treatment. Plast Reconstr Surg. 2008;121:258–68.
16. Rohrich RJ, Pessa JE. The fat compartments of the face: anatomy and clinical implications for cosmetic surgery. Plast Reconstr Surg. 2007;119(7):2219–27. https://doi.org/10.1097/01.prs.0000265403.66886.54. PMID: 17519724.
17. Kohn LA. The molecular basis of skin aging. Front Bioeng Biotechnol. 2019;7:59.
18. Carruthers J, et al. A multicenter, randomized, double-blind, placebo-controlled study of the efficacy and safety of IncobotulinumtoxinA in the treatment of glabellar frown lines using a validated 6-point Photonumeric scale. Dermatol Surg. 2017;43(9):1141–50.
19. Sherrington CS. The integrative action of the nervous system. Yale University Press; 1947.
20. Mendelson B, Wong CH. Changes in the facial skeleton with aging: implications and clinical applications in facial rejuvenation. Aesthet Plast Surg. 2012;36(4):753–60. https://doi.org/10.1007/s00266-012-9904-3. Epub 2012 May 12. PMID: 22580543; PMCID: PMC3404279.
21. Gosain AK, Klein MH, Sudhakar PV, Prost RW. A volumetric analysis of soft-tissue changes in the aging midface using high-resolution MRI: implications for facial rejuvenation. Plast Reconstr Surg. 2005;115(4):1143–52; discussion 1153-5. https://doi.org/10.1097/01.prs.0000156333.57852.2f. PMID: 15793457.
22. Vierkötter A, Schikowski T, Ranft U, et al. Airborne particle exposure and extrinsic skin aging. J Invest Dermatol. 2010;130(12):2719–26.
23. Li M, Vierkötter A, Schikowski T, et al. Epidemiological evidence that indoor air pollution from cooking with solid fuels accelerates skin aging in Chinese women. J Dermatol Sci. 2015;79(2):148–54.
24. El-Domyati M, Attia S, Saleh F, et al. Intrinsic aging vs. photoaging: a comparative histopathological, immunohistochemical, and ultrastructural study of skin. Exp Dermatol. 2002;11(5):398–405.
25. Naylor EC, Watson RE, Sherratt MJ. Molecular aspects of skin ageing. Maturitas. 2011;69(3):249–56.
26. Langton AK, Sherratt MJ, Griffiths CE, Watson RE. A new wrinkle on old skin: the role of elastic fibres in skin ageing. Int J Cosmet Sci. 2010;32(5):330–9.
27. Uitto J. The role of elastin and collagen in cutaneous aging: intrinsic aging versus photoexposure. J Drugs Dermatol. 2008;7(2 Suppl):s12–6.
28. Farkas JP, Pessa JE, Hubbard B, Rohrich RJ. The science and theory behind facial aging. Plast Reconstr Surg Glob Open 2013;1(1):e8-e15. doi:https://doi.org/10.1097/GOX.0b013e31828ed1da. PMID: 25289202; PMCID: PMC4174174.

29. Makrantonaki E, Zouboulis CC. Molecular mechanisms of skin aging: state of the art. Ann N Y Acad Sci. 2007;1119:40–50. https://doi.org/10.1196/annals.1404.027. PMID: 18056953.
30. Campisi J, d'Adda di Fagagna F. Cellular senescence: when bad things happen to good cells. Nat Rev Mol Cell Biol. 2007;8(9):729–40. https://doi.org/10.1038/nrm2233. PMID: 17667954.
31. Rittié L, Fisher GJ. UV-light-induced signal cascades and skin aging. Ageing Res Rev. 2002;1(4):705–20. https://doi.org/10.1016/s1568-1637(02)00024-7. PMID: 12208239.
32. Poljšak B, Dahmane R. Free radicals and extrinsic skin aging. Dermatol Res Pract. 2012;2012:135206. https://doi.org/10.1155/2012/135206. Epub 2012 Feb 29. PMID: 22505880; PMCID: PMC3299230
33. Thornton MJ. Estrogens and aging skin. Dermatoendocrinol. 2013;5(2):264–70. https://doi.org/10.4161/derm.23872. PMID: 24194966; PMCID: PMC3772914
34. Victorelli S, Passos JF. Telomeres and cell senescence - size matters not. EBioMedicine. 2017;21:14–20. https://doi.org/10.1016/j.ebiom.2017.03.027. Epub 2017 Mar 21. PMID: 28347656; PMCID: PMC5514392.
35. Coleman SR, Katzel EB. Fat grafting for facial filling and regeneration. Clin Plast Surg. 2015;42(3):289–300, vii. https://doi.org/10.1016/j.cps.2015.04.001. PMID: 26116934.
36. Charles-de-Sá L, Gontijo-de-Amorim NF, Maeda Takiya C, Borojevic R, Benati D, Bernardi P, Sbarbati A, Rigotti G. Antiaging treatment of the facial skin by fat graft and adipose-derived stem cells. Plast Reconstr Surg. 2015;135(4):999–1009.
37. Zuk PA, Zhu M, Mizuno H, Huang J, Futrell JW, Katz AJ, Benhaim P, Lorenz HP, Hedrick MH. Multilineage cells from human adipose tissue: implications for cell-based therapies. Tissue Eng. 2001;7(2):211–28. https://doi.org/10.1089/107632701300062859. PMID: 11304456.
38. Surowiecka A, Chrapusta A, Klimeczek-Chrapusta M, Korzeniowski T, Drukała J, Strużyna J. Mesenchymal stem cells in burn wound management. Int J Mol Sci. 2022;23(23):15339. https://doi.org/10.3390/ijms232315339. PMID: 36499664; PMCID: PMC9737138.
39. Van den Broek LJ, Kroeze KL, Waaijman T, Breetveld M, Sampat-Sardjoepersad SC, Niessen FB, Middelkoop E, Scheper RJ, Gibbs S. Differential response of human adipose tissue-derived mesenchymal stem cells, dermal fibroblasts, and keratinocytes to burn wound exudates: potential role of skin-specific chemokine CCL27. Tissue Eng Part A. 2014;20(1–2):197–209. https://doi.org/10.1089/ten.TEA.2013.0123. Epub 2013 Oct 2. PMID: 23980822; PMCID: PMC3875150.
40. Butterwick KJ, Bevin AA, Iyer S. Fat grafting to the periorbital region and cheek. J Drugs Dermatol. 2007;6(3):309–12.
41. Rihani J. Microfat and Nanofat: when and where these treatments work. Facial Plast Surg Clin North Am. 2019;27(3):321–30. https://doi.org/10.1016/j.fsc.2019.03.004. PMID: 31280846.
42. Kamat P, Frueh FS, McLuckie M, Sanchez-Macedo N, Wolint P, Lindenblatt N, Plock JA, Calcagni M, Buschmann J. Adipose tissue and the vascularization of biomaterials: stem cells, microvascular fragments and nanofat-a review. Cytotherapy. 2020;22(8):400–11. https://doi.org/10.1016/j.jcyt.2020.03.433. Epub 2020 Jun 2. PMID: 32507607.
43. Cicione C, Vadalà G, Di Giacomo G, Tilotta V, Ambrosio L, Russo F, Zampogna B, Cannata F, Papalia R, Denaro V. Micro-fragmented and nanofat adipose tissue derivatives: in vitro qualitative and quantitative analysis. Front Bioeng Biotechnol. 2023;11:911600. https://doi.org/10.3389/fbioe.2023.911600. PMID: 36733959; PMCID: PMC9887143.
44. Tonnard P, Verpaele A, Peeters G, Hamdi M, Cornelissen M, Declercq H. Nanofat grafting: basic research and clinical applications. Plast Reconstr Surg. 2013;132(4):1017–26. https://doi.org/10.1097/PRS.0b013e31829fe1b0. PMID: 23783059.
45. Memar O, Nezamabadi A, Milani BY, Milani FY, Djalilian A. Nanofat grafting: basic research and clinical application. Plast Reconstr Surg. 2014;133(5):728e. https://doi.org/10.1097/PRS.0000000000000135. PMID: 24776592.
46. Strong AL, Rohrich RJ, Tonnard PL, Vargo JD, Cederna PS. Technical precision with autologous fat grafting for facial rejuvenation: a review of the evolving science. Plast Reconstr Surg. 2023; https://doi.org/10.1097/PRS.0000000000010643. Epub ahead of print. PMID: 37159906.

47. Bernardini FP, Gennai A, Izzo L, Zambelli A, Repaci E, Baldelli I, Fraternali-Orcioni G, Hartstein ME, Santi PL, Quarto R. Superficial enhanced fluid fat injection (SEFFI) to correct volume defects and skin aging of the face and periocular region. Aesthet Surg J. 2015;35(5):504–15. https://doi.org/10.1093/asj/sjv001. Epub 2015 Apr 24. PMID: 25911629.
48. Riesco B, Abascal C, Duarte A, Flores RM, Rouaux G, Sampayo R, Bernardini F, Devoto M. Autologous fat transfer with SEFFI (superficial enhanced fluid fat injection) technique in periocular reconstruction. Orbit. 2018;37(3):191–5. https://doi.org/10.1080/01676830.2017.1383470. Epub 2017 Oct 17. PMID: 29040029.
49. Gennai A, Zambelli A, Repaci E, Quarto R, Baldelli I, Fraternali G, Bernardini FP. Skin rejuvenation and volume enhancement with the micro superficial enhanced fluid fat injection (M-SEFFI) for skin aging of the periocular and perioral regions. Aesthet Surg J. 2017;37(1):14–23. https://doi.org/10.1093/asj/sjw084. Epub 2016 May 30. PMID: 27241362.
50. Gennai A, Bernardini FP, Baldessin M, Bovani B, Camporese A, et al. Evaluation of the viability and Phenotipe of adipose derived cells harvested using different harvesting and processing procedures: a pilot study. J stem cells clin pract. 2022;2(1):101.
51. Alessandro Gennai, Bovani B, Colli M, Melfa F, Piccolo D, Russo R, Barbara Roda, Andrea Zattoni, Pierluigi Reschiglian, Silvia Zia. Comparison of harvesting and processing technique for adipose tissue graft: evaluation of cell viability. Int J Regen Med. 2021:2613–5914. https://doi.org/10.31487/j.RGM.2021.02.03.
52. Coleman SR, Grover R. The anatomy of the aging face: volume loss and changes in 3-dimensional topography. Aesthet Surg J. 2006;26(1S):S4–9. https://doi.org/10.1016/j.asj.2005.09.012. PMID: 19338976.
53. Wysong A, Joseph T, Kim D, Tang JY, Gladstone HB. Quantifying soft tissue loss in facial aging: a study in women using magnetic resonance imaging. Dermatologic Surg. 2013;39(12):1895–902. https://doi.org/10.1111/dsu.12362. Epub 2013 Nov 14. PMID: 24238002.
54. Plowes Hernandez O, Montes Bracchini JJ. Management of the Heavy Brows: long-term surgical options. Facial Plast Surg. 2018;34(1):36–42. https://doi.org/10.1055/s-0037-1617422. Epub 2018 Feb 6. PMID: 29409102.
55. Fitzgerald R. Contemporary concepts in brow and eyelid aging. Clin Plast Surg. 2013;40(1):21–42. https://doi.org/10.1016/j.cps.2012.08.005. PMID: 23186754.
56. Odunze M, Rosenberg DS, Few JW. Periorbital aging and ethnic considerations: a focus on the lateral canthal complex. Plast Reconstr Surg. 2008;121(3):1002–8.
57. Camp MC, Wong WW, Filip Z, Carter CS, Gupta SC. A quantitative analysis of periorbital aging with three-dimensional surface imaging. J Plast Reconstr Aesthet Surg. 2011;64(2):148–54.
58. Friedman O. Changes associated with the aging face. Facial Plast Surg Clin North Am. 2005;13(3):371–80. https://doi.org/10.1016/j.fsc.2005.04.004. PMID: 16085282.
59. Kane MA, Cox SE, Jones D, Lei X, Gallagher CJ. Heterogeneity of crow's feet line patterns in clinical trial subjects. Dermatologic Surg. 2015;41(4):447–56. https://doi.org/10.1097/DSS.0000000000000336. PMID: 25789814.
60. Alghoul M, Codner MA. Retaining ligaments of the face: review of anatomy and clinical applications. Aesthet Surg J. 2013;33(6):769–82. https://doi.org/10.1177/1090820X13495405. Epub 2013 Jul 12. PMID: 23855010.
61. Rohrich RJ, Hollier LH Jr, Janis JE, Kim J. Rhinoplasty with advancing age. Plast Reconstr Surg. 2004;114(7):1936–44. https://doi.org/10.1097/01.prs.0000143308.48146.0a. PMID: 15577372.
62. Rohrich RJ, Pessa JE, Ristow B. The youthful cheek and the deep medial fat compartment. Plast Reconstr Surg. 2008;121(6):2107–12. https://doi.org/10.1097/PRS.0b013e31817123c6. PMID: 18520902.
63. Pecora NG, Baccetti T, McNamara JA Jr. The aging craniofacial complex: a longitudinal cephalometric study from late adolescence to late adulthood. Am J Orthod Dentofacial Orthop. 2008;134(4):496–505. https://doi.org/10.1016/j.ajodo.2006.11.022. PMID: 18929267.
64. Vashi NA, de Castro Maymone MB, Kundu RV. Aging differences in ethnic skin. J Clin Aesthet Dermatol. 2016;9(1):31–8. PMID: 26962390; PMCID: PMC4756870

65. Jacovella PF. Use of calcium hydroxylapatite (Radiesse) for facial augmentation. Clin Interv Aging. 2008;3:161–74.
66. Berlin AL, Hussain M, Goldberg DJ. Calcium hydroxylapatite filler for facial rejuvenation: a histologic andimmunohistochemical analysis. Dermatologic Surg. 2008;34(suppl 1):S64–7.
67. DeLorenzi C. Complications of injectable fillers, part I: vascular complications. Aesthet Surg J. 2014;34(4):561–75.
68. Tansatit T, et al. Risk zones of necrosis after filler injection: Angiosome and beyond. Plast Reconstr Surg Glob Open. 2018;6(4):e1724.
69. Beleznay K, et al. Vascular complications of dermal fillers: a review of the literature. Dermatol Surg. 2015;41(9):1097–117.
70. Qian H, Ling Y, Zhang M, Lenahan C, Wang C, Zheng Z, Shao A, Zhang J. Front Hum Neurosci. 2021. Sec. Brain Imaging and Stimulation Volume 15–2021; https://doi.org/10.3389/fnhum.2021.610945.
71. Sito G, Manzoni V, Sommariva R. Vascular complications after facial filler injection: a literature review and meta-analysis. J Clin Aesthet Dermatol. 2019;12(6):E65–72. Epub 2019 Jun 1. PMID: 31360292; PMCID: PMC6624005
72. Gennai A, Baldessin M, Melfa F, Bovani B, Camporese A, Claysset B, Colli M, Diaspro A, Russo R, Strano P, Bollero D, Capparè G, Casadei A, Gallo G, Piccolo D, Salti G, Tesauro P. Guided superficial enhanced fluid fat injection (SEFFI) procedures for facial rejuvenation: an Italian multicenter retrospective case report. Clin Pract. 2023;13(4):924–43. https://doi.org/10.3390/clinpract13040085. PMID: 37623266; PMCID: PMC10453478.
73. Schiraldi L, Sapino G, Meuli J, Maruccia M, Cherubino M, Raffoul W, di Summa PG. Facial fat grafting (FFG): worth the risk? A systematic review of complications and critical appraisal. J Clin Med. 2022;11:4708.
74. Urdiales-Gálvez F, Delgado NE, Figueiredo V, Lajo-Plaza JV, Mira M, Moreno A, Ortíz-Martí F, Del Rio-Reyes R, Romero-Álvarez N, Del Cueto SR, Segurado MA, Rebenaque CV. Treatment of Soft Tissue Filler Complications: Expert Consensus Recommendations. Aesthet Plast Surg. 2018;42(2):498–510. https://doi.org/10.1007/s00266-017-1063-0. Epub 2018 Jan 5. PMID: 29305643; PMCID: PMC5840246.
75. Wang K, Rong X, Dang J, Yang J, Zheng H, Hou M, Li H, Jiang C, Xiong S, Qiu L, Yu Z, Yi C. Severe vascular complications caused by facial autologous fat grafting: a critical review. Ann Plast Surg. 2021;86(3S Suppl 2):S208–19. https://doi.org/10.1097/SAP.0000000000002691. PMID: 33443884.
76. Chatrath V, Banerjee PS, Goodman GJ, Rahman E. Soft-tissue filler-associated blindness: a systematic review of case reports and case series. Plast Reconstr Surg Glob Open. 2019;7(4):e2173. https://doi.org/10.1097/GOX.0000000000002173. PMID: 31321177; PMCID: PMC6554164.
77. Kim YK, Jung C, Woo SJ, et al. Cerebral angiographic findings of cosmetic facial filler-related ophthalmic and retinal artery occlusion. J Korean Med Sci. 2015;30:1847.
78. Clauser L, Lucchi A, Tocco-Tussardi I, Gardin C, Zavan B. Autologous fat transfer for facial augmentation and regeneration: role of mesenchymal stem cells. Atlas of the Oral and Maxillofacial Surgery Clinics. 2018;26:25–32.

Chapter 3
Three-Dimensional Medical Facelift with Autologous Regenerative Therapy and Calcium Hydroxyapatite

3.1 Introduction

Skin aging is a multifaceted process influenced by both intrinsic (genetic) and extrinsic (environmental) factors. While intrinsic aging is an inherent aspect driven by genetic predispositions, extrinsic aging is largely a result of external influences such as exposure to UV radiation, pollution, and lifestyle choices [9, 22, 30].

Both forms of aging contribute to the deterioration of the extracellular matrix (ECM), a critical network of macromolecules essential for maintaining skin structure and function [10].

The ECM comprises various components including collagen, elastin, fibronectin, and hyaluronic acid, each playing a distinct role in preserving skin texture and firmness. Fibroblasts, specialized cells within the dermis, are responsible for producing collagen and elastin, which provide structural support and elasticity to the skin, respectively. Fibronectin aids in ECM regeneration, while hyaluronic acid ensures adequate hydration levels. Any alterations in these components can lead to visible signs of aging such as wrinkles, fine lines, and loss of skin elasticity [20].

In addition to intrinsic and extrinsic factors, the persistence of senescent fibroblasts in tissues further exacerbates the aging process. These cells release senescence-associated secretory phenotype (SASP), creating a microenvironment conducive to aging by activating matrix metalloproteinases (MMPs) that degrade the ECM and suppress the release of growth factors. Moreover, the dysregulation of MMPs, enzymes crucial for maintaining ECM balance, is deeply implicated in skin aging. In youthful skin, MMP activity is tightly regulated to preserve ECM integrity. However, with aging, this balance is disrupted, leading to excessive MMP activity that accelerates ECM degradation, resulting in the formation of wrinkles, sagging, and other visible signs of aging [23].

Understanding these intricate mechanisms underlying skin aging is essential for developing effective antiaging strategies. Regenerative aesthetic treatments aim to

D. Piccolo et al., *A.R.T. Autologous Regenerative Therapy in Aesthetic Medicine*, https://doi.org/10.1007/978-3-032-05835-5_3

address age-related tissue changes by targeting cellular signaling and improving the tissue microenvironment. By promoting angiogenesis and enhancing ECM remodelling, these treatments help to restore skin health and vitality, ultimately combating the signs of aging.

As time progresses, usually starting around the age of 25, fibroblasts gradually decrease their activity, leading to a decline in the production of collagen and elastin. Initially, expression lines become noticeable, followed by skin laxity around the age of 40, and eventually, deeper wrinkles, known as gravitational wrinkles, start to develop after the age of 50 [11].

Although this aging process is natural and inevitable, it can be partially prevented and treated. Prevention involves maintaining a healthy lifestyle, which, while not offering miraculous results, helps sustain good skin quality. In terms of treatment, aesthetic medicine offers various minimally invasive options. Over the past two decades, facelift techniques have undergone significant evolution. To achieve a natural-looking rejuvenation, it is crucial to acknowledge the three-dimensional nature of the face. Therefore, adopting a three-dimensional facelift approach, which comprehensively addresses the face, is crucial for achieving a balanced and natural rejuvenation. This approach may involve treatments like dermal fillers and Autologous Regenerative Therapy (ART). Possessing considerable regenerative potential, these treatments can effectively mitigate the signs of aging and partially restore a youthful appearance. Overall, they hold significant promise in rejuvenating the face and restoring a more youthful look [26].

3.2 Dermal Fillers

Dermal fillers are injectable treatments that utilize molecules with biostimulating effects. When administered beneath the skin, these molecules stimulate the dermis to generate new collagen molecules, replenishing the reserves lost over time. It is important to note that biostimulation is a gradual process, while the results are not immediately visible, they improve over time as the body gradually produces new collagen.

One of the most appealing aspects of biostimulation treatments, besides being noninvasive or minimally invasive, is the naturalness of the results. The improvement in the appearance of the skin tissue and the rejuvenating effect are always in harmony with the individual as a whole. What is achieved is a reduction, or at least a significant attenuation, of wrinkles along with a lifting of the skin tissue, resulting in a redefinition of the jawline or, more generally, the contour of the face [31].

3.2.1 Hyaluronic Acid Fillers

In particular, the utilization of hyaluronic acid (HA) fillers in aesthetic medicine has evolved significantly over the years, emerging as one of the most prevalent procedures worldwide. HA fillers are among the most popular because they have a soft consistency that creates natural-looking results. HA is naturally found in the body, so it naturally and gradually dissolves over time, typically between 6 and 12 months depending on the individual and the product used. The main concepts of HA filler usage include rejuvenation, enhancement, and restoration. Typically, HA fillers have a duration of effectiveness ranging between 4 and 24 months. This duration is influenced by factors such as cross-linking, particle size, gel state (monophasic or biphasic), injection volume and location, as well as individual host metabolism.

3.2.2 Calcium Hydroxyapatite Fillers

Calcium hydroxyapatite (CaHA) filler, commercially available as CaHA (Merz aesthetics, Frankfurt, Germany), has been approved by the US Food and Drug Administration (FDA) in 2006 for the treatment of facial wrinkles and folds and subsequently approved in 2009 for more aesthetic indications. Treatment areas on label include nasolabial folds, marionette lines, periorbital region, medial and lateral cheek, midface, jawline angle, chin, jawline, and rejuvenation of the dorsal aspect of the hands [1].

CaHA is a biodegradable synthetic compound composed of microspheres containing calcium and phosphate. The microspheres measure from 20 to 45 μm in diameter and are suspended in a 70% aqueous gel matrix including glycerin (6.4%) and sodium carboxymethylcellulose (1.3%), which dissolves within 1 to 3 months after injection [32].

CaHA is a substance naturally found in the human body as a physiological inorganic constituent of bones and teeth. This characteristic renders the product fully biocompatible, biodegradable, non-mutagenic, non-antigenic, and highly safe both locally and systemically. When injected, CaHA initially causes mechanical tissue expansion primarily due to the volume of the carrier gel. However, the microspheres within CaHA induce an inflammatory response, leading to biostimulation and fibroplasia, which typically begins around 1 month after injection. This process ultimately results in collagen synthesis. Therefore, the volume observed immediately after CaHA injection is largely due to the carrier gel, which dissolves within the first 3 months. Subsequently, the volume is maintained by the collagen biostimulation.

CaHA fillers typically last between 10 and 14 months, influenced by factors such as the injected volume and the individual's metabolism [18].

CaHA possesses a unique property that gives CaHA the ability to cause less swelling compared to HA fillers: due to its innovative composition, CaHA does not absorb water, making it less hygroscopic, and consequently, it does not induce swelling after injection. On the contrary, HA fillers are hygroscopic, meaning they attract water molecules, resulting in a turgor effect [24].

3.2.2.1 Mechanism of Action

CaHA stands out as the only dermal device with a well-defined dual mechanism of action. In its undiluted form, it offers immediate volumization followed by long-lasting collagen biostimulation. However, when volumization is not the primary objective, diluted or hyperdiluted forms of CaHA can bypass the initial volume-enhancing effect. This allows immediate utilization of the product's biostimulation properties. Diluted and hyperdiluted CaHA can thus be tailored to specific patient categories seeking to manage laxity and enhance skin quality and firmness. These formulations provide a compacting effect on tissues through the biostimulation of neocollagenesis and angiogenesis [24].

Various studies have explored the mechanism of action of CaHA fillers (Table 3.1). Yutskovskaya et al. compared HA and CaHA fillers in stimulating neocollagenesis and assessing dermal elastin levels at 4- and 9-month post-administration. Dermal collagen, constituting 77% of skin dry weight, primarily consists of type I and III collagen, with type I being crucial for skin strength and resilience. In neocollagenesis, type III collagen is gradually replaced by type I collagen, resulting in tissue with high tensile strength. CaHA stimulates neocollagenesis more effectively than HA, promoting a two-phase process of type III collagen production followed by type I collagen, replacing the former. Yutskovskaya's study suggests that the CaHA gel matrix consistently remodels ECM in line with this process. At 4 months, type III collagen staining was significantly higher with CaHA gel compared to HA gel. By 9 months, type I collagen staining was higher with CaHA gel, while type III collagen staining was lower, indicating effective remodeling.

Moreover, the CaHA gel matrix notably stimulated cell proliferation in comparison to HA gel, thereby supporting enhanced collagen production. Furthermore, the CaHA gel matrix facilitated neoangiogenesis at 4 and 9 months, promoting increased blood flow and nutrient delivery. This effect is likely linked to the early synthesis of type III collagen, underscoring CaHA's beneficial impact on skin rejuvenation and remodeling [33].

Additional research by Yutskovskaya has highlighted that CaHA treatment provides immediate correction followed by a gradual tissue regeneration process involving neocollagenesis, elastin production, dermal cell proliferation, and angiogenesis. This complete approach leads to long-lasting aesthetic improvements, such as firmer, more elastic skin and increased thickness. Unlike certain HA fillers that induce neocollagenesis through mechanical stretching, CaHA stimulates targeted collagen biostimulation in the injection area. This is facilitated by CaHA microspheres, which act as scaffolds for new tissue formation and activate fibroblasts,

Table 3.1 List of main studies assessing biological changes after CaHA administration

Biological effects	Time post-administration (months)	Number of patients	Reference
Collagen and elastin formation	4 and 9 months	24	[33]
Collagen formation	6 months	5	[2]
Collagen formation	6 months	16	[21]
Elastin formation	6 months	15	[17]
Collagen and elastin formation	4 and 7 months	20	[34]

resulting in collagen and elastin production. Furthermore, injecting diluted CaHA into the neck and décolleté skin has demonstrated stimulation of collagen and elastin synthesis, along with increased neovascularization. This remodeling process is associated with enhanced dermal thickness and improved mechanical properties of the skin, including elasticity and flexibility [34].

Additionally, a nonhuman study on dogs also showed an increase in host collagen following intradermal injections of CaHA over a period ranging from 4 to 78 weeks [6].

In a human study involving five subjects aged 55 to 73 years exhibiting prominent nasolabial rhytids, it was observed that the CaHA filler stimulates the deposition of new collagen and infiltration around the injected material at 6 months. This phenomenon likely plays a role in the clinical improvement observed in treated wrinkles. While type I collagen is primarily expressed at the 6-month mark, lower amounts of type III collagen may also be present in the infiltration into the new stroma [2].

Moreover, Kim and colleagues studied how to improve soft tissue in atrophic ischial areas with CaHA filler injections. Sixteen women received injections and visible improvements in skin quality and roughness were observed in all patients. Multilayered injections effectively treated atrophic ischial skin with no significant side effects [21].

It has been found that the CaHA gel matrix also stimulates ECM remodeling by increasing elastin levels to a significantly greater extent compared to HA gel. Indeed, elastin staining intensity was significantly higher after treatment with CaHA gel matrix vs HA gel, both at month 4 and month 9 after treatment. Elastin provides the skin with the ability to recoil after experiencing deformative stresses. Therefore, unlike HA gel, the CaHA gel matrix appears to have the potential to enhance not only the structural resilience of the skin through collagen remodeling but also its elasticity, properties that are known to diminish with skin aging in parallel with the loss of both collagen and elastin fibers [33].

In particular, in another human study involving healthy female subjects aged 35 to 60, Gonzáles et al. examined the response of injected CaHA in human skin, particularly focusing on its ability to induce the formation of new elastin and proteoglycans. This investigation utilized histological analysis of skin samples obtained from volunteers who had received this filler. This study demonstrated an increase in proteoglycans following the injection of CaHA. Proteoglycans play a crucial role in providing hydration and tissue swelling pressure, enabling the skin to resist compression forces. Therefore, CaHA contributes to the structural support of tissues. Regarding changes in elastin at 6 months, it is worth noting a distinction from the findings of Yutskovskaya and colleagues. They consistently demonstrated a significant increase in elastin at the 4-month mark in a study population notably younger, ranging from 35 to 45 years of age. This observation aligns with the clinical finding that neocollagenesis induced by fillers, and presumably new elastin and proteoglycan formation, is more prominently observed in younger individuals [17].

3.2.2.2 Procedures and Safety

In 2009, the FDA approved a protocol to mix CaHA with lidocaine to enhance patient comfort during injection. In 2016, CaHA (+) Lidocaine, containing integrated 0.3% lidocaine, also received Conformité Européenne (CE) marking and is now available for physicians in Europe. Diluted CaHA has been widely used by injectors as it broadens the range of indications for which CaHA can be used. The FDA-approved mixing protocol to produce a final concentration of 0.3% lidocaine represents the optimal protocol to preserve the viscoelastic properties of CaHA when used as a volumizing filler. However, many experts have developed techniques to use higher dilutions of CaHA to harness its collagen-stimulating effects for reducing skin laxity without the volumizing effect. Dilutions can be titrated based on the thickness of the patient's skin and the degree of tissue laxity to ensure easy product placement [34].

Recent research has delved into the effects of different dilution ratios on CaHA dermal fillers, aiming to optimize dispersion and concentration of CaHA microspheres for enhanced aesthetic outcomes. Indeed, it was demonstrated that CaHA contact with fibroblasts was a key factor for the induction of neocollagenesis and targeted dilution may improve this contact in vivo since more dilute CaHA resulted in a greater dissemination of CaHA microspheres compared to less dilute solutions. It was found that a 1:2 hyperdilution of CaHA resulted in increased dispersion and decreased concentration of CaHA microspheres compared to a 1:1 dilution. This suggests that altering the dilution ratio can impact the distribution and density of microspheres at the injection site, potentially influencing treatment outcomes. Further study is needed to optimize dilution ratios for CaHA fillers, balancing dispersion, concentration, and clinical efficacy for enhanced aesthetic results [27].

Global consensus guidelines have defined recommendations for the dilution of CaHA. These guidelines emphasize performing the dilution immediately before use in order to maintain product integrity. It is essential to create a sterile mixing environment to ensure the safety of the procedure. To facilitate the dilution process, practitioners are advised to use a 3–10 mL counter syringe, with larger syringes recommended for higher dilutions. This choice of syringe size allows for greater flexibility and precision in achieving the desired concentration of the CaHA solution. The correct dilution procedure involves connecting a LuerLock syringe containing the diluent to the original CaHA syringe through a female transfer adapter. Both syringes should be sufficiently large to easily accommodate the combined filler and diluent, particularly when working with higher dilutions and larger total volumes. This ensures that the mixture is thoroughly blended and homogenous. To achieve optimal product homogeneity, it is recommended to pass the mixture between the syringes at least 20 times. This meticulous approach helps to evenly distribute the components and ensures consistency in the final CaHA solution. By adhering to these guidelines, practitioners can ensure the safe and effective dilution of CaHA, maximizing the potential benefits for patients while minimizing the risk of complications [15].

A recent systemic review on the properties of CaHA analyzed the degree of effectiveness and safety of the product, concluding that the use of CaHA can be considered a first-choice treatment for volumizing and lifting effects, superior to existing alternatives. CaHA is generally safe when used with caution. Complications can arise from injection-related events, suboptimal technique, or true complications. Injection-related events include bruising, swelling, redness, and pain, common with most injectable treatments. Suboptimal technique may lead to papules and nodules, which can resolve spontaneously. True complications like granulomas are rare but may require corticosteroid or 5-fluorouracil injections. Inflammatory nodules may indicate infection, necessitating antibiotic treatment. Vascular complications are rare and may cause tissue necrosis or, rarely, blindness. CaHA should not be injected superficially or in highly mobile areas like the lips, where muscle movement can lead to nodule formation. Additionally, it is contraindicated for glabellar wrinkle correction. When injecting highly vascularized areas, it is crucial to exercise caution and use small quantities [1].

3.3 Autologous Regenerative Therapy (ART)

Adipose tissue, abundant in mesenchymal stem cells (MSCs), has garnered attention for its regenerative potential, particularly within the stromal vascular fraction (SVF) [8].

The SVF is a diverse mix of cells, including pericytes, smooth muscle cells, and notably, adipose-derived stem cells (ADSCs), which are essential for tissue regeneration and repair. ADSCs, in particular, possess multipotent capabilities, allowing them to transform into various cell types like adipocytes, osteoblasts, and chondrocytes. Moreover, they release angiogenic and anti-inflammatory factors, contributing to tissue healing. The combined action of these cell types, along with immune cells present in SVF, enhances its therapeutic efficacy. The accessibility and high stem cell content of SVF have sparked significant research interest, driving investigations into its regenerative potential across different fields. The SVF, a diverse cellular mixture obtainable from adipose tissue, offers extensive regenerative potential that can be utilized for skin rejuvenation and tissue regeneration. These cells assist in skin regeneration, produce vital components of the ECM, and possess beneficial anti-inflammatory properties for skin quality [16, 29].

ADSCs possess specific characteristics such as phenotype markers (negative for CD45, CD235a, CD31, and positive for CD34, CD90, CD73, CD105, and CD44), ability to attach and expand on plastic culture plates, and multipotent differentiation potential into adipocytes, chondrocytes, and osteoblasts under suitable conditions. ADSCs primarily exert their therapeutic effects through the secretion of a regenerative secretome rather than direct cellular differentiation. This secretome contains essential components like ECM proteins (collagens, laminin, fibronectin) and various MMPs crucial for the healing process. ADSCs also play an immunomodulatory role, promoting inflammation in non-inflammatory environments and inhibiting inflammation in the presence of inflammatory factors. In chronic wounds where inflammation is blocked, ADSCs may primarily act through their anti-inflammatory functions rather than their pro-inflammatory/angiogenic activity.

Additionally, ADSCs stimulate angiogenesis through the secretion of angiogenic factors like angiopoietins, vascular endothelial growth factor, fibroblast growth factor-2, platelet-derived growth factor, transforming growth factor-β, and hepatocyte growth factor, as well as support the proliferation of dermal fibroblasts and keratinocytes, leading to dermo-epidermal reconstruction. They also produce antifibrotic molecules, express pattern recognition receptors against pathogens, and exhibit antimicrobial activity. While the multipotent capabilities of ADSCs may contribute to their therapeutic action, their ability to differentiate into dermal fibroblasts or keratinocyte-like cells may accelerate the formation of granulation tissue and promote epidermal regeneration. ADSCs are typically found in the SVF of adipose tissue, which can be directly utilized for therapeutic interventions, although many studies utilize purified in vitro autologous or allogeneic ADSCs or their secreted exosomes [4, 29].

3.4 Harvesting Procedure (Superficial Enhanced Fluid Fat Injection, SEFFI)

The restoration of facial volume is often achieved through autologous methods such as fat transfer. Indeed, autologous adipose tissue has been described as the ideal filler, which provides a natural filling in a safe and simple procedure [7].

However, this technique has limitations such as unpredictable fat survival and a significant risk of visible lumps. Consequently, many researchers have focused their efforts on microfat grafting. Indeed, the injection of small (0.2 to 0.8 mm) clusters of adipose tissue improves cellular blood irrigation, along with the degree of engraftment [12].

Recently, a new technique named Superficial Enhanced Fluid Fat Injection (SEFFI) has been developed. This novel approach involves the use of micro cannulas with lateral ports to extract small quantities of fat, eliminating the need for additional postharvest processing [3].

Indeed, harvesting superficial adipose tissue enables the collection of tissue that is richer in SVF cells. These cells play a crucial role in tissue regeneration processes [5, 28].

This harvested fat, abundant in stem cells and crucial adipocytes, is subsequently enriched and rendered more fluid through the addition of autologous platelet-rich plasma (PRP). Following this enhancement process, the fat is meticulously injected into the target area using syringe needles, ensuring precise application and optimal results, representing a significant advancement in fat grafting procedures [3, 12].

3.5 Combination of CAHA with Guided SEFFI Procedure

Recent advancements in aesthetic medicine have sparked interest in the synergistic effects of combining adipose tissue implantation with dermal fillers for enhanced skin rejuvenation. Particularly, investigations into the compatibility of adipose tissue and dermal fillers, such as CaHA, have shown promising results, opening avenues for novel regenerative aesthetic approaches. In our recent study, micro-fragmented fatty tissue (FAT) grafting was combined with the CaHA filler procedure to assess its biological impact on FAT. Samples of micro-fragmented adipose tissue were harvested from five patients using the guided SEFFI technique [14].

3.5.1 Procedure

In summary, adipose tissue was extracted from either the abdomen or the trochanteric region. Manual aspiration of fat tissue was carried out using a 10 mL syringe equipped with a plunger lock and a microperforated side port cannula inserted into a patented guide with side port holes of 800 microns. Aspiration involved gentle back-and-forth fan movements across the sampling area, ensuring subcutaneous tunneling adjacent to the dermis in the superficial adipose tissue (SAT). Collection ceased after obtaining 5 mL of tissue, which was then rinsed with saline solution and allowed to stratify into two layers by gravity for a few minutes. After removing the washing liquid, the tissue was placed into five disposable sterile plastic boxes without coding and promptly transported to Tecnopolo Mirandola "Mario Veronesi" via "Croce Blu" transfer in a dedicated refrigerated container. Upon arrival, each sample was assigned an internal conventional number code, and ten syringe vials containing 1.5 mL each of CaHA filler were provided at room temperature.

The combination of lipoaspirates and CaHA followed the protocol outlined in a previous study on lipoaspirates and HA [14].

Fat tissue was collected in a 30 mL syringe and washed three times with 15 mL of phosphate-buffered saline (PBS) to remove blood cells. Two different culture conditions were established: "FAT" and "FAT + R". In detail:

- FAT: 0.5 mL of adipose tissue (AT) was seeded onto each well of MW6 (MW6 Cellstar Greine Bio-one), prefilled with 2 mL of culture medium aMEM (MEM Alpha Medium), and enriched with 2.5% PLP (Cell Culture-Human Platelet Lysate), 0.2% heparin, 1% glutamine, and 0.5% ciprofloxacin.
- FAT + R: A mixture of 0.6 mL of AT and CaHA in a ratio of 1 (CaHA):4 (lipoaspirate) was prepared. CaHA was collected using a syringe connector, mixed until homogeneously dispersed in the filler, and then seeded.

The specimens were incubated at 37 °C, 5% CO_2 for 4 days to allow the release and adhesion of stromal cells to the plastic surface.

After the initial 4 days, the culture media were changed and refreshed every 2–3 days, ensuring not to disturb the floating AT fragment.

Proliferation, differentiation, and immunophenotype assays were then performed [25].

3.5.2 *Biological Effects*

Our recent study examined the biological impact of combining CaHA dermal filler with fat obtained through the guided SEFFI procedure. Specifically, our findings indicate that the combination of FAT and CaHA is biocompatible. We observed consistent proliferation rates among isolated cells, with similar population doubling times and numbers of doubling cells in both the FAT and FAT + R groups. Furthermore, immunophenotypic analysis using fluorescence-activated cell sorting revealed that adherent cells from both culture conditions (FAT and FAT + R) exhibited specific markers for mesenchymal stem/stromal cells [25].

In addition, the same study evaluated multipotential differentiation using differentiation assays. Results showed that in both conditions (FAT and FAT + R), the adipogenic, chondrogenic, and the osteogenic potential was well preserved [25].

These findings underscore the compatibility and regenerative potential of this combination for facial antiaging treatments.

Furthermore, a retrospective study has explored the safety and efficacy of combining SEFFI with CaHA dermal filler for cosmetic enhancements. These studies revealed a low complication rate, with ecchymosis being the most commonly reported adverse event. The collective evidence from these studies supports the notion that the combination of SEFFI and CaHA is safe and offers favorable outcomes in aesthetic procedures [13].

Another retrospective observational study has been conducted in patients treated with SEFFI in combination with CaHA from January 2021 to March 2023, and results will be published soon.

These findings collectively highlight the promising potential of combining adipose tissue implantation with CaHA dermal fillers for facial rejuvenation. Not only does this approach appear to be biocompatible, but it also offers a safe and effective strategy for addressing facial aging concerns. As research in this area continues to evolve, it is expected to further refine and optimize aesthetic procedures, ultimately benefiting patients seeking natural and long-lasting improvements in skin quality and appearance.

3.5.3 *Facial Injection Procedure*

The combination of guided SEFFI alongside CaHA has demonstrated remarkable natural outcomes in rejuvenating the face, with effects that endure over time. The injection procedure occurs in two stages. Initially, CaHA is injected either in a 1:1

Table 3.2 List of injection sites and the related CAHA dilutions and administration

Area	CAHA dilution	Cannula	Injection plan
Temples/eyebrow arch	1:1 with saline solution	25G and/or 27G	Subdermal, supraperiosteal
Malar/zigomatic area	1:1 with saline solution	25G and/or 27G	Supraperiosteal, subcutaneous
Preauricular area/ mandibular angle	1:1 with saline solution	25G and/or 27G	Subcutaneous
Cheeks	1:1 or 1:2 with saline solution	25G and/or 27G	Subcutaneous, subdermal
Chin/oral commissure/ marionette	1:1 with saline solution	25G and/or 27G	Subdermal, supraperiosteal

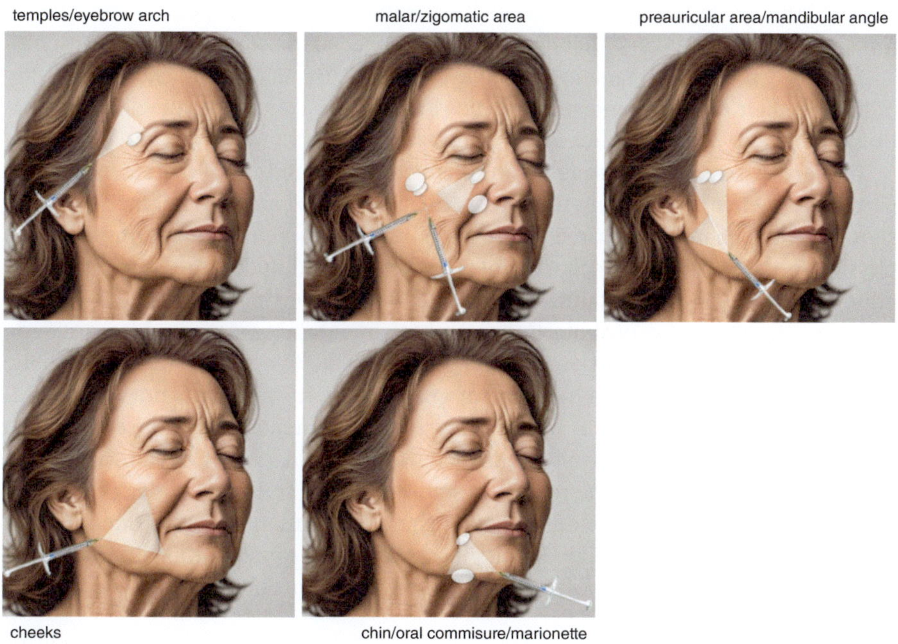

Fig. 3.1 Labeled facial injection sites shown across five procedural images

dilution ratio or with hyperdilution up to 1:2 with saline solution, tailored to the specific injection sites such as the temporal, malar, zygomatic, preauricular, mandibular angle, and marionette areas (Table 3.2 and Fig. 3.1). This is administered using 38 mm 27 G or 50 mm 25 G cannulas, targeting the supraperiosteal and subcutaneous or subdermal planes (Table 3.2). Subsequently, on the same day or 2 weeks later, the guided SEFFI treatment is executed, involving the implantation of approximately 20–24 mL of previously collected adipose tissue. This is accurately placed within the subcutaneous and subdermal planes, utilizing 25 G or 22 G 50 mm cannulas.

3.6 Patient Assessment

Facial aging is a multifactorial and three-dimensional process involving loss of volume and changes in the skin. Every anatomical component of the face is affected, including the bones, ligaments, soft tissues, fat, and skin. This aging process varies among individuals due to genetic and anatomical disparities, with some exhibiting signs of midfacial aging as early as their third decade while others may not experience changes until their sixth decade. Key components of midfacial aging include the descent of the soft tissue envelope, leading to the development of tear-trough deformities and nasolabial folds, as well as devolumization of midface structures [19].

Bone mass and fat volume tend to decrease, while muscles become more flaccid, ligaments lose tension, and the epidermal layer of the skin becomes loose (Fig. 3.2).

Any strategy in aesthetic medicine must refer to the assessment of the state of impairment of the layers that make up the face: skin, subcutaneous, musculo-aponeurotic, retaining ligaments and space, periosteum, and deep fascia. The study of tissues, their mobility, and structure has made it possible to adapt facial rejuvenation products to achieve increasingly natural results.

In facial assessment, distinguishing between fixed and mobile areas is crucial, particularly when considering the effects of aging. Fixed areas of the face typically show minimal change over time and play an important role in maintaining a unique physiognomy. Serving as anchor points for the skin and underlying muscles, these regions provide essential stability. Moreover, the supporting ligaments associated with these fixed areas contribute significantly to maintaining structural integrity and providing additional support during aging. Among the two facial areas, the mobile region is more susceptible to the force of gravity. Tissue mobility is notably heightened in the central facial area, where adipose compartments gradually separate and displace. This increased mobility contributes to the formation of dynamic wrinkles and folds, contrasting with the relatively stable nature of fixed facial areas (Fig. 3.3).

Understanding the differential effects of gravity on mobile versus fixed regions is pivotal in developing effective antiaging strategies, as interventions may need to address both structural support and dynamic tissue movement to achieve optimal rejuvenation outcomes.

As the human face ages, it undergoes a complex transformation marked by the sagging and shifting of mobile areas. These changes, driven by factors like collagen

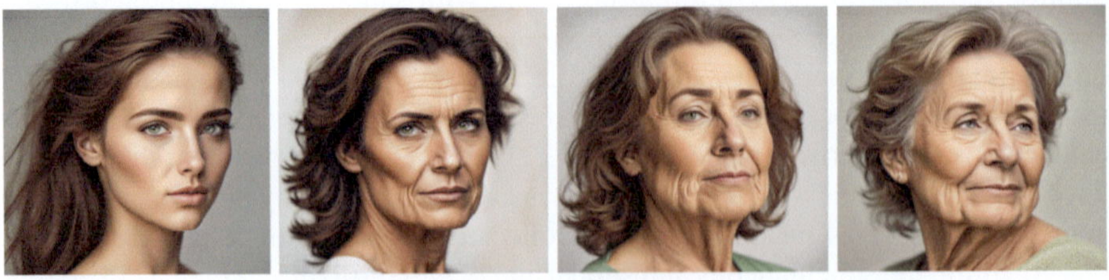

Fig. 3.2 Portrait series showing aging progression through facial feature changes

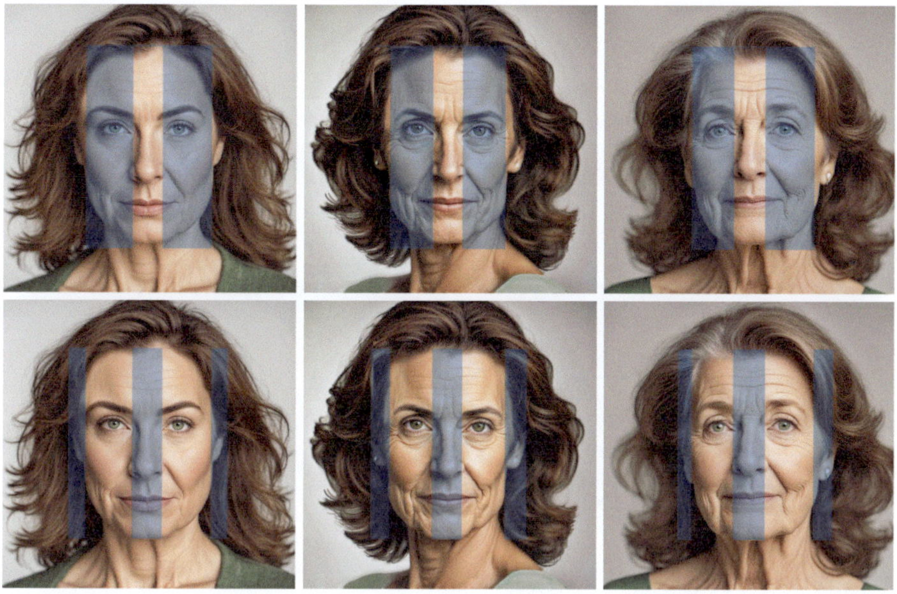

Fig. 3.3 Facial region mobility changes illustrated across different age groups

Fig. 3.4 Progressive facial aging shown through split-face comparison images

depletion and gravity, manifest in predictable patterns across individuals. Despite the uniqueness of each face, certain zones consistently exhibit signs of sagging and tissue slippage (Fig. 3.4).

In addition, assessing skin thickness across various facial regions is a critical consideration in understanding facial aesthetics and addressing cosmetic concerns. Thin skin often found around delicate areas like the eyes and lips tends to be more translucent.

On the other hand, thick skin, typically seen on the forehead and chin, presents challenges due to its robust nature, requiring tailored approaches for effective treatment. Medium skin, which encompasses the majority of facial surfaces, strikes a balance between these extremes (Fig. 3.5).

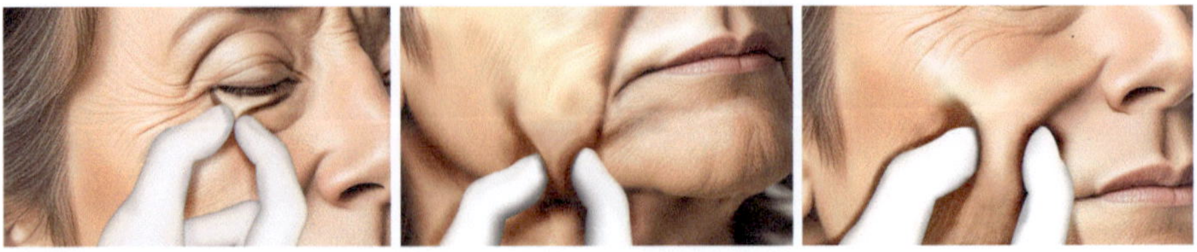

Fig. 3.5 Skin thickness variations illustrated with corresponding wrinkle visibility

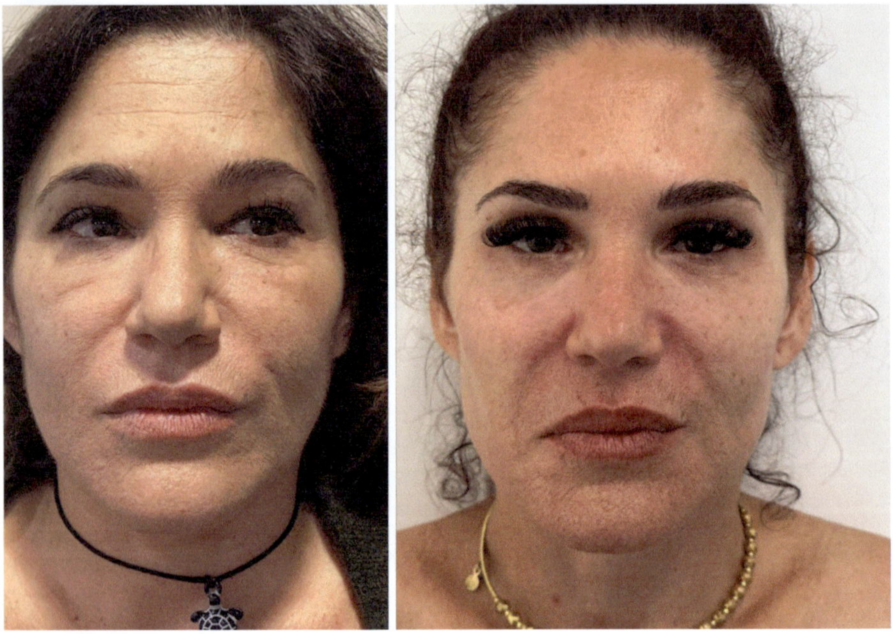

Fig. 3.6 Before and after images showing facial enhancements post-SEFFI procedure

By understanding and respecting the unique characteristics of thin, medium, and thick skin, practitioners can direct the complexities of facial aesthetics with precision, which is crucial for achieving a balanced and natural rejuvenation, enhancing patient satisfaction and safety.

We firmly believe that by conducting a meticulous assessment of patients' characteristics and employing the advanced techniques outlined here, we can provide personalized care at any age, from the first signs of aging onward, ensuring harmonious and natural outcomes (Figs. 3.6, 3.7, 3.8, 3.9, 3.10, 3.11, 3.12, 3.13, 3.14, 3.15, and 3.16).

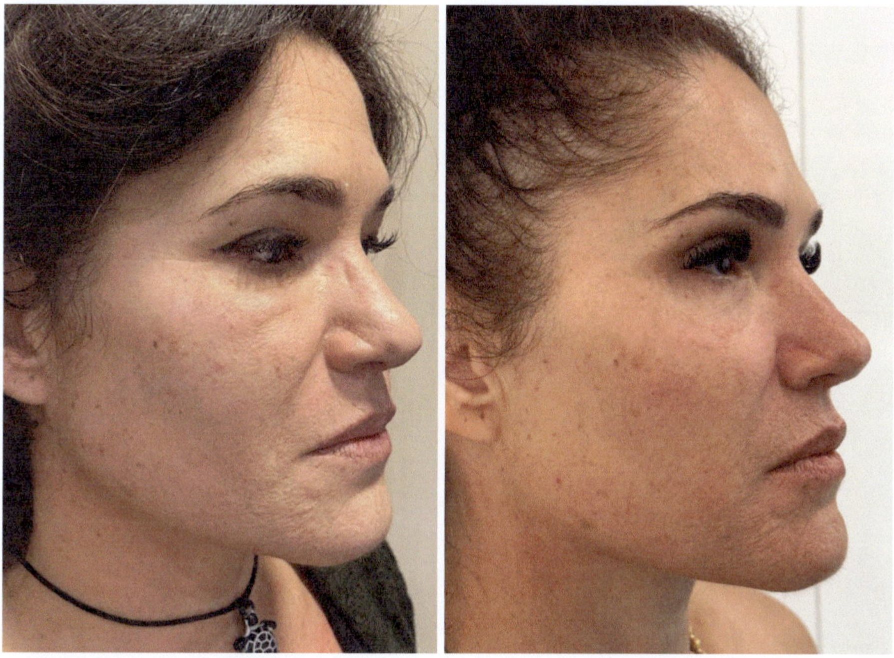

Fig. 3.7 Before and after comparison highlighting improved skin and jawline definition

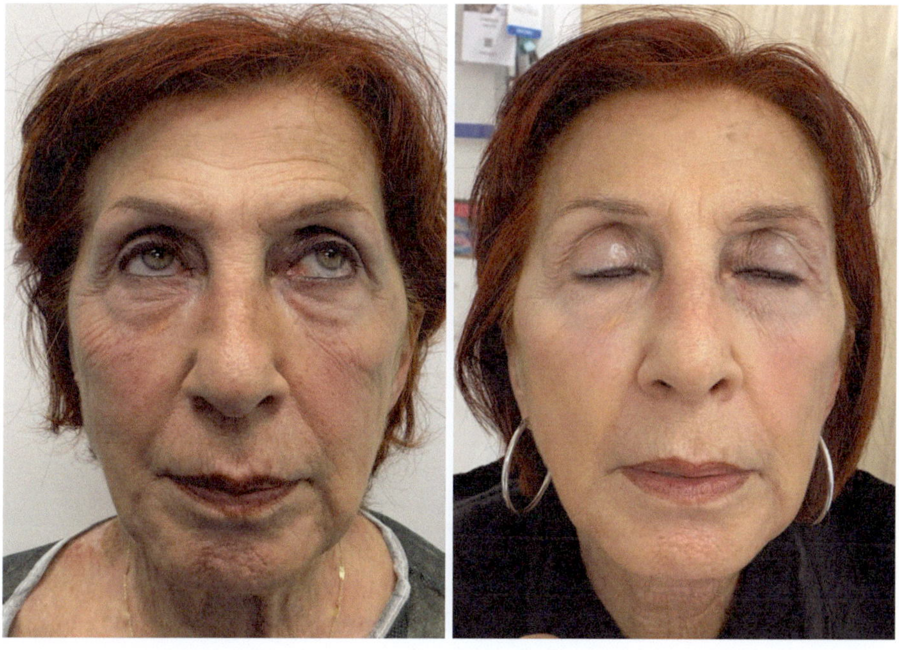

Fig. 3.8 Before and after images showing reduced under-eye bags and wrinkles

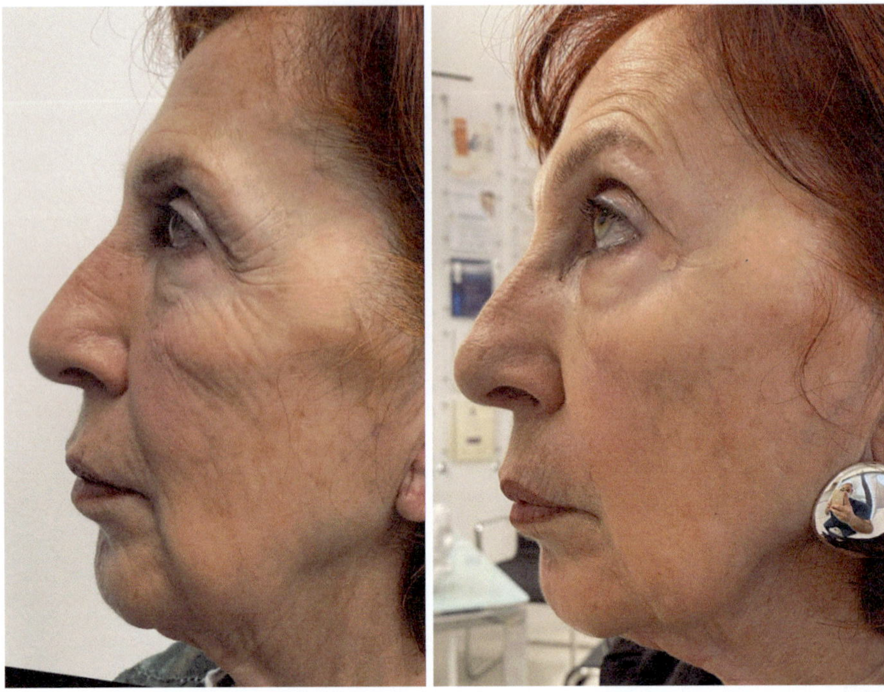

Fig. 3.9 Before and after comparison showing reduced wrinkles and rejuvenation

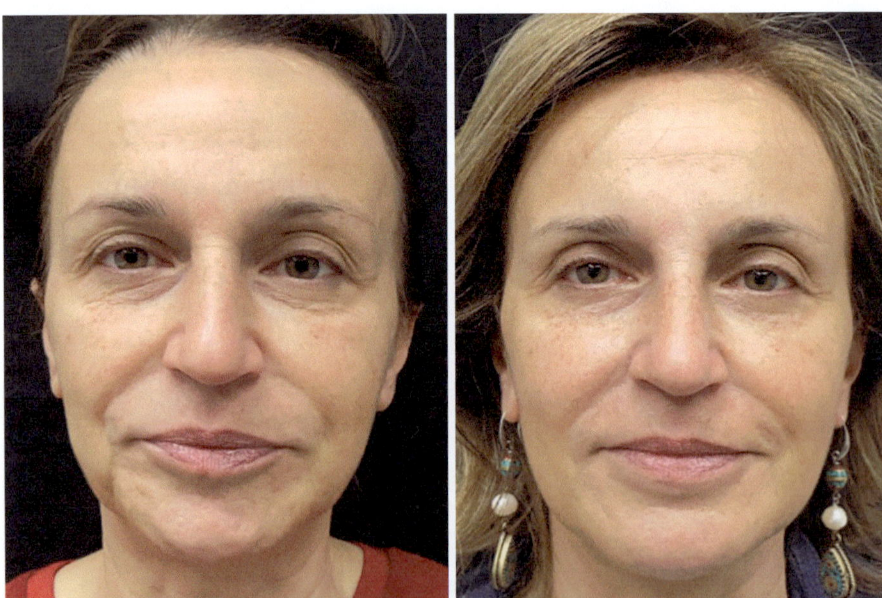

Fig. 3.10 Before and after images showing subtle facial enhancements post-procedure

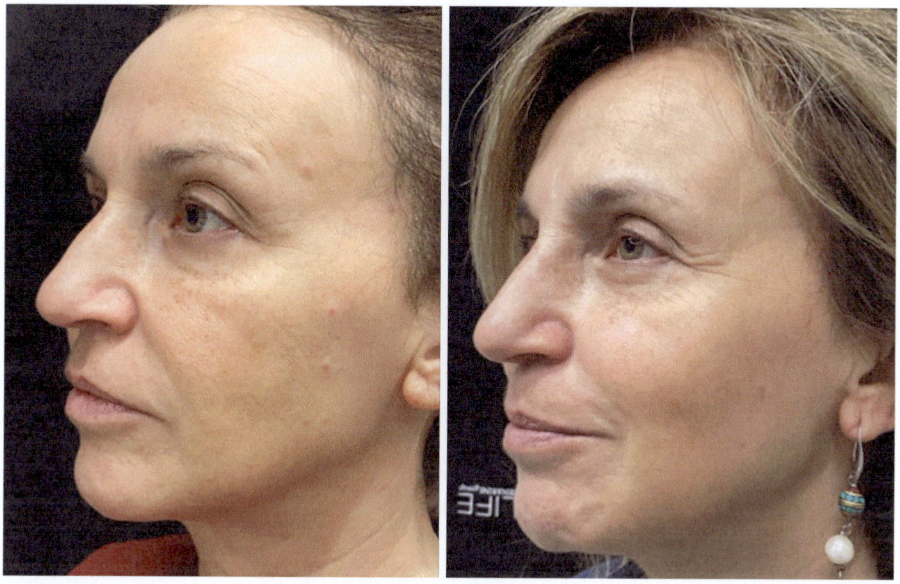

Fig. 3.11 Before and after comparison showing smoother skin and improved complexion

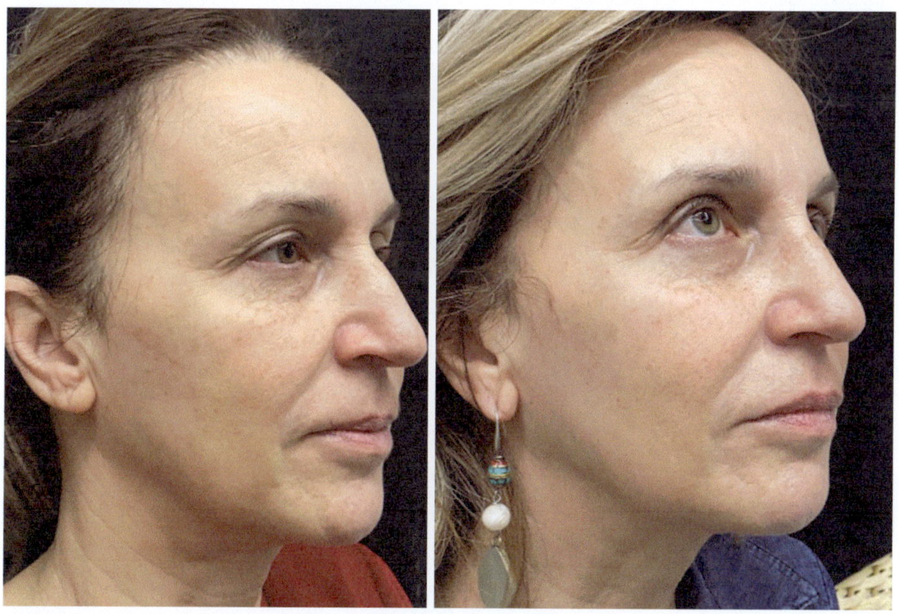

Fig. 3.12 Before and after images highlighting smoother skin and youthful appearance

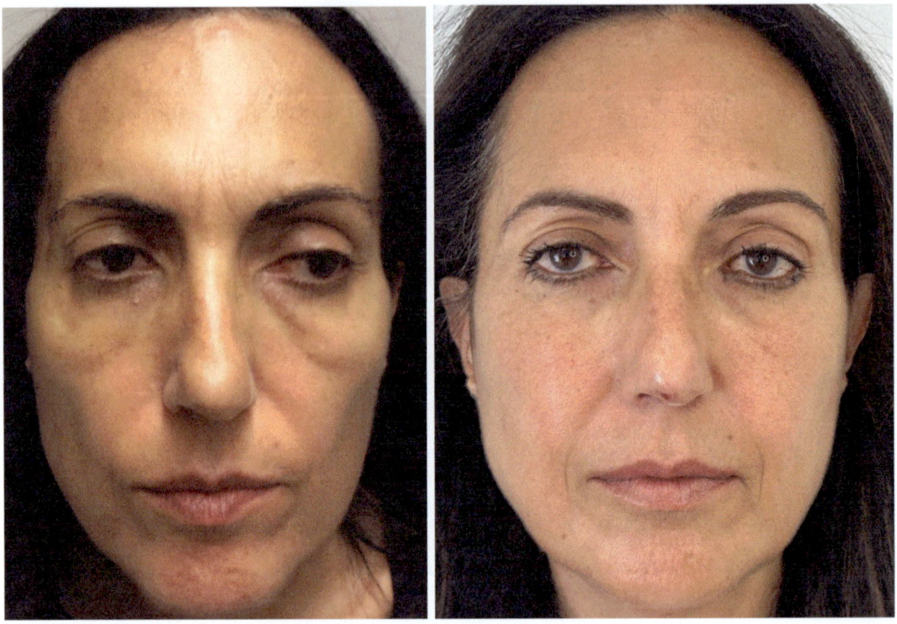

Fig. 3.13 Before and after showing reduced under-eye circles and smoother complexion

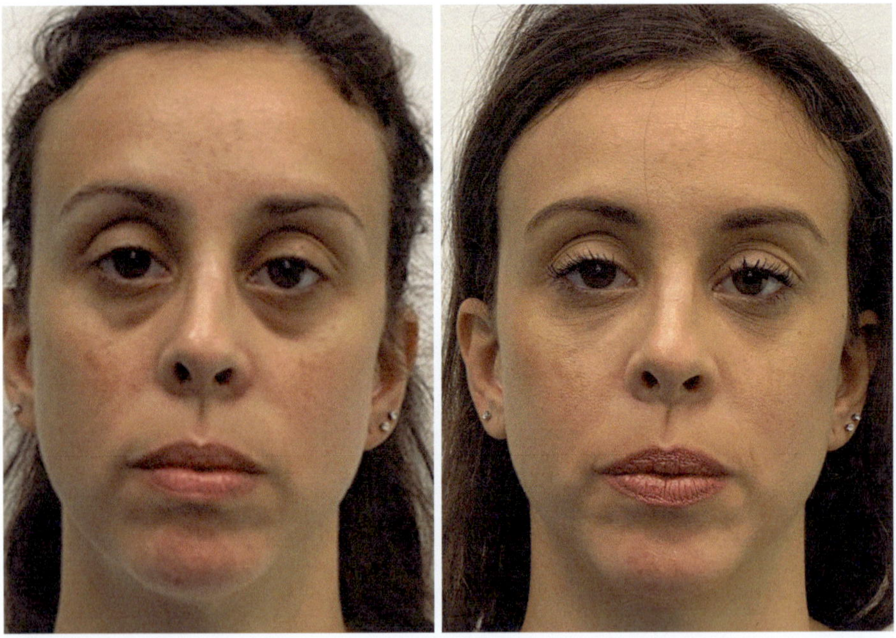

Fig. 3.14 Before and after images showing smoother skin and subtle enhancements

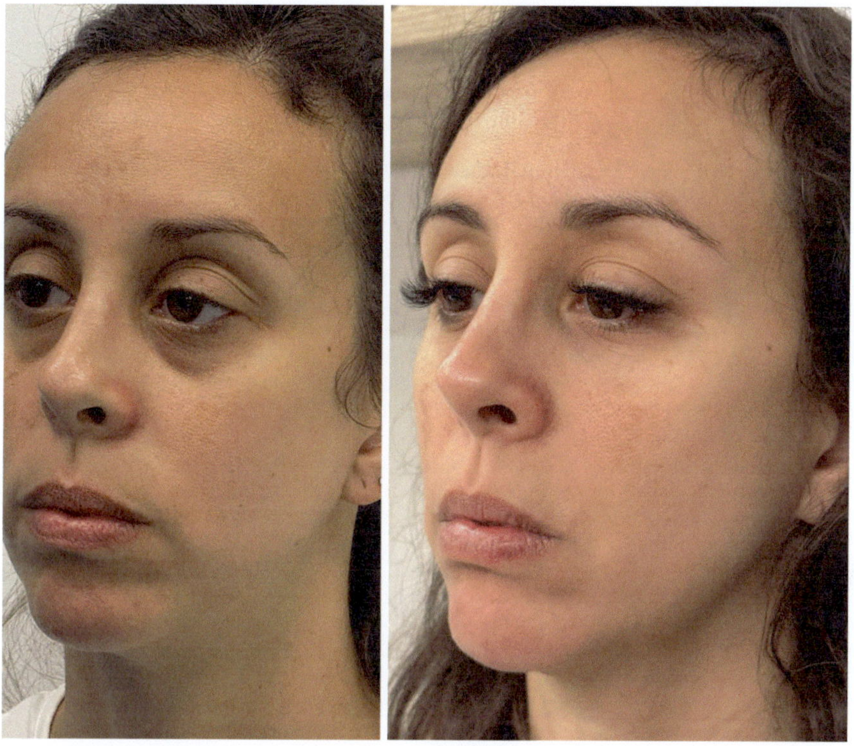

Fig. 3.15 Before and after comparison showing smoother skin and subtle smile

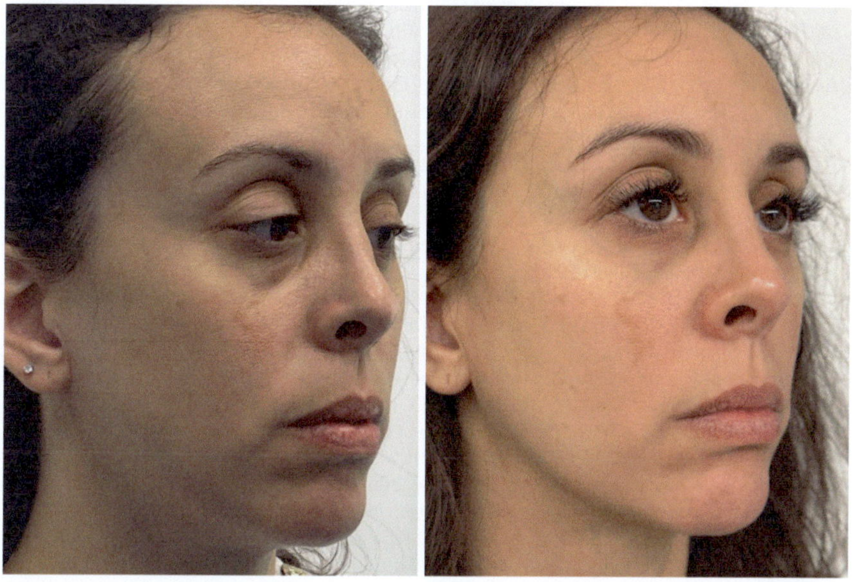

Fig. 3.16 Before and after images showing smoother skin and enhanced facial definition

References

1. Bartoletti E, Melfa F, Renzi M, Rovatti P. Systematic review of the literature on the properties, quality and reliability of calcium hydroxyapatite: results of an Italian experts' meeting. Aesthetic Med. 2022;8(1):11–24.
2. Berlin AL, Hussain M, Goldberg DJ. Calcium hydroxylapatite filler for facial rejuvenation: a histologic and immunohistochemical analysis. Dermatologic Surg. 2008;34(Suppl 1):S64–7. https://doi.org/10.1111/j.1524-4725.2008.34245.x. PMID: 18547184.
3. Bernardini FP, Gennai A, Izzo L, Zambelli A, Repaci E, Baldelli I, Fraternali-Orcioni G, Hartstein ME, Santi PL, Quarto R. Superficial enhanced fluid fat injection (SEFFI) to correct volume defects and skin aging of the face and periocular region. Aesthet Surg J. 2015;35(5):504–15. https://doi.org/10.1093/asj/sjv001. Epub 2015 Apr 24. PMID: 25911629.
4. Brembilla NC, Vuagnat H, Boehncke WH, Krause KH, Preynat-Seauve O. Adipose-derived stromal cells for chronic wounds: scientific evidence and roadmap toward clinical practice. Stem Cells Transl Med. 2023;12(1):17–25. https://doi.org/10.1093/stcltm/szac081. PMID: 36571216; PMCID: PMC9887085.
5. Cappellano G, Morandi EM, Rainer J, Grubwieser P, Heinz K, Wolfram D, Bernhard D, Lobenwein S, Pierer G, Ploner C. Human macrophages preferentially infiltrate the superficial adipose tissue. Int J Mol Sci. 2018;19(5):1404. https://doi.org/10.3390/ijms19051404. PMID: 29738484; PMCID: PMC5983635.
6. Coleman KM, Voigts R, DeVore DP, Termin P, Coleman WP III. Neocollagenesis after injection of calcium hydroxylapatite composition in a canine model. Dermatol Surg. 2008;34(Suppl 1):S53–5. https://doi.org/10.1111/j.1524-4725.2008.34243.x. PMID: 18547182.
7. Coleman SR. Facial recontouring with lipostructure. Clin Plast Surg. 1997;24(2):347–67. PMID: 9142473.
8. Crisan M, Yap S, Casteilla L, Chen CW, Corselli M, Park TS, Andriolo G, Sun B, Zheng B, Zhang L, Norotte C, Teng PN, Traas J, Schugar R, Deasy BM, Badylak S, Buhring HJ, Giacobino JP, Lazzari L, Huard J, Péault B. A perivascular origin for mesenchymal stem cells in multiple human organs. Cell Stem Cell. 2008;3(3):301–13. https://doi.org/10.1016/j.stem.2008.07.003. PMID: 18786417.
9. El-Domyati M, Attia S, Saleh F, et al. Intrinsic aging vs. photoaging: a comparative histopathological, immunohistochemical, and ultrastructural study of skin. Exp Dermatol. 2002;11(5):398–405. https://doi.org/10.1034/j.1600-0625.2002.110502.x. PMID: 12366692.
10. Estrach S, Vivier CM, Féral CC. ECM and epithelial stem cells: the scaffold of destiny. Front Cell Dev Biol. 2024;12:1359585. https://doi.org/10.3389/fcell.2024.1359585. PMID: 38572486; PMCID: PMC10987781.
11. Fujimura T, Haketa K, Hotta M, Kitahara T. Loss of skin elasticity precedes to rapid increase of wrinkle levels. J Dermatol Sci. 2007;47(3):233–9. https://doi.org/10.1016/j.jdermsci.2007.05.002. Epub 2007 Jun 18. PMID: 17576050.
12. Gennai A, Bernardini FP. Superficial enhanced fluid fat injection (SEFFI and MicroSEFFI) in facial rejuvenation. CellR4. 2017;5(1):e2239.
13. Gennai A, Baldessin M, Melfa F, Bovani B, Camporese A, Claysset B, Colli M, Diaspro A, Russo R, Strano P, Bollero D, Capparè G, Casadei A, Gallo G, Piccolo D, Salti G, Tesauro P. Guided superficial enhanced fluid fat injection (SEFFI) procedures for facial rejuvenation: an Italian multicenter retrospective case report. Clin Pract. 2023;13(4):924–43. https://doi.org/10.3390/clinpract13040085. PMID: 37623266; PMCID: PMC10453478.
14. Gennai A, Bovani B, Colli M, Melfa F, Piccolo D, Russo R, Tretti Clementoni M, Zia S, Roda B, Zattoni A. Evaluation of the Number, Biophysical and Multipotent Characteristics of Adipose Derived Stem Cells Harvested by SEFFI Procedure and Interaction with Different Type of Hyaluronic Acids. Int J Regen Med. 2021; https://doi.org/10.31487/j.RGM.2021.02.02.
15. Goldie K, Peeters W, Alghoul M, Butterwick K, Casabona G, Chao YYY, Costa J, Eviatar J, Fabi SG, Lupo M, Sattler G, Waldorf H, Yutskovskaya Y, Lorenc P. Global consensus guidelines for the injection of diluted and Hyperdiluted calcium Hydroxylapatite for skin tightening.

Dermatol Surg. 2018;44(Suppl 1):S32–41. https://doi.org/10.1097/DSS.0000000000001685. Erratum in: Dermatologic Surg 2019;45(2):327. PMID: 30358631.

16. Goncharov EN, Koval OA, Igorevich EI, Encarnacion Ramirez MJ, Nurmukhametov R, Valentinovich KK, Montemurro N. Analyzing the clinical potential of stromal vascular fraction: a comprehensive literature review. Medicina (Kaunas). 2024;60(2):221. https://doi.org/10.3390/medicina60020221. s
17. González N, Goldberg DJ. Evaluating the effects of injected calcium Hydroxylapatite on changes in human skin elastin and proteoglycan formation. Dermatologic Surg. 2019;45(4):547–51. https://doi.org/10.1097/DSS.0000000000001809. PMID: 30893178.
18. Herrmann JL, Hoffmann RK, Ward CE, Schulman JM, Grekin RC. Biochemistry, physiology, and tissue interactions of contemporary biodegradable injectable dermal fillers. Dermatologic Surg. 2018;44(Suppl 1):S19–31. https://doi.org/10.1097/DSS.0000000000001582. PMID: 29994947.
19. Jacono AA, Rousso JJ. An algorithmic approach to multimodality midfacial rejuvenation using a new classification system for midfacial aging. Clin Plast Surg. 2015;42(1):17–32. https://doi.org/10.1016/j.cps.2014.08.002. PMID: 25440738.
20. Jariwala N, Ozols M, Bell M, Bradley E, Gilmore A, Debelle L, Sherratt MJ. Matrikines as mediators of tissue remodelling. Adv Drug Deliv Rev. 2022;185:114240. https://doi.org/10.1016/j.addr.2022.114240. Epub 2022 Apr 1. PMID: 35378216.
21. Kim J. Multilayered injection of calcium hydroxylapatite filler on ischial soft tissue to rejuvenate the previous phase of chronic sitting pressure sore. Clin Cosmet Investig Dermatol. 2019;12:771–84. https://doi.org/10.2147/CCID.S212599. PMID: 31686889; PMCID: PMC6800547.
22. Li M, Vierkötter A, Schikowski T, et al. Epidemiological evidence that indoor air pollution from cooking with solid fuels accelerates skin aging in Chinese women. J Dermatol Sci. 2015;79(2):148–54. https://doi.org/10.1016/j.jdermsci.2015.04.001. PMID: 26055797.
23. Low E, Alimohammadiha G, Smith LA, Costello LF, Przyborski SA, von Zglinicki T, Miwa S. How good is the evidence that cellular senescence causes skin ageing? Ageing Res Rev. 2021;71:101456. https://doi.org/10.1016/j.arr.2021.101456. Epub 2021 Sep 3. PMID: 34487917; PMCID: PMC8524668
24. Massidda E. Starting point for protocols on the use of Hyperdiluted calcium Hydroxylapatite (CaHA) for optimizing age-related biostimulation and rejuvenation of face, neck, décolletage and hands: a case series report. Clin Cosmet Investig Dermatol. 2023;16:3427–39. https://doi.org/10.2147/CCID.S420068. PMID: 38050476; PMCID: PMC10693750.
25. Melfa F, Gennai A, Carfagna G, Bovani B, Piccolo D, Colli M, Baldessin M, Siragusa D. Characterization of adipose-derived mesenchymal stem cells from tissue harvested with the guided SEFFI technique and co-cultured with calcium hydroxyapatite. J Appl Cosmetol. 2023;41(2):4–19. https://doi.org/10.56609/jac.v41i2.288.
26. Mirrafati S, Shiffman M. Three-dimensional facelift. In: Shiffman M, Mirrafati S, Lam S, Cueteaux C, editors. Simplified facial rejuvenation. Berlin/Heidelberg: Springer; 2008. https://doi.org/10.1007/978-3-540-71097-4_50.
27. Nowag B, Casabona G, Kippenberger S, Zöller N, Hengl T. Calcium hydroxylapatite microspheres activate fibroblasts through direct contact to stimulate neocollagenesis. J Cosmet Dermatol. 2023;22(2):426–32. https://doi.org/10.1111/jocd.15521. Epub 2022 Dec 27. PMID: 36575882.
28. Rossi M, Roda B, Zia S, Vigliotta I, Zannini C, Alviano F, Bonsi L, Zattoni A, Reschiglian P, Gennai A. Characterization of the tissue and stromal cell components of micro-superficial enhanced fluid fat injection (micro-SEFFI) for facial aging treatment. Aesthet Surg J. 2020;40(6):679–90. https://doi.org/10.1093/asj/sjy142. PMID: 29905790.
29. Tallone T, Realini C, Böhmler A, Kornfeld C, Vassalli G, Moccetti T, Bardelli S, Soldati G. Adult human adipose tissue contains several types of multipotent cells. J Cardiovasc Transl Res. 2011;4(2):200–10. https://doi.org/10.1007/s12265-011-9257-3. Epub 2011 Feb 15. PMID: 21327755.

30. Vierkötter A, Schikowski T, Ranft U, et al. Airborne particle exposure and extrinsic skin aging. J Invest Dermatol. 2010;130(12):2719–26. https://doi.org/10.1038/jid.2010.204. PMID: 20664556.
31. Wanick FBF, Issa MCA, Pontello R, Tamura B. Fillers. In: Issa M, Tamura B, editors. Daily routine in cosmetic dermatology. Clinical approaches and procedures in cosmetic dermatology, vol 1. Cham: Springer; 2017. https://doi.org/10.1007/978-3-319-12589-3_33.
32. Yag-Howard C, DeNigris J. Novel filler technique: hyaluronic acid and calcium hydroxylapetite mixture resulting in favorable esthetic and longevity outcomes. Int J Womens Dermatol. 2021;7(5 Pt B):817–9. https://doi.org/10.1016/j.ijwd.2021.09.008. PMID: 35028387; PMCID: PMC8714586.
33. Yutskovskaya Y, Kogan E, Leshunov E. A randomized, split-face, histomorphologic study comparing a volumetric calcium hydroxylapatite and a hyaluronic acid-based dermal filler. J Drugs Dermatol. 2014;13(9):1047–52. PMID: 25226004.
34. Yutskovskaya YA, Kogan EA. Improved Neocollagenesis and Skin Mechanical Properties After Injection of Diluted Calcium Hydroxylapatite in the Neck and Décolletage:A Pilot Study. J Drugs Dermatol. 2017;16(1):68–74. PMID: 28095536.

Chapter 4
Energy-Based Devices and Autologous Regenerative Therapy: A Synergy with Great Potential

4.1 Premise

Facial skin aging is a complex process that fully involves all present tissues. Skin quality progressively declines, with loss of luminosity and tone, followed by a change in volume of the subcutaneous layer. Some adipose compartments, both deep and superficial, become impoverished while others become hypertrophic. This is further accentuated by progressive hypotrophy of the underlying muscular and bony tissues and the loss of tone in the present ligamentous structures. This progressive process, which affects all present anatomical structures, ultimately results in a progressive downward sliding of the cutaneous and subcutaneous tissues, drawn by gravity. The resulting tissue laxity is the primary source of loss of definition in the jawline, nasolabial and marionette folds, as well as the submental and neck regions. Patients commonly seek to regain the aesthetic characteristics of earlier years without compromising their overall appearance, ideally through natural-looking results. The clinical objective, therefore, is to restore lost volume and skin contours, enhancing tissue brightness and firmness. Autologous Regenerative Therapy will represent an indispensable weapon, thanks to its original characteristics: no heterologous substance used, no risk of adverse reactions, great regenerative potential, unique in its kind, with the added possibility of reharmonizing lost volumes by varying the degree of fluidification of the adipose tissue to be implanted according to needs. A key question is whether adequate restoration of tissue tone can be achieved using Autologous Regenerative Therapy alone. An important consideration is how to effectively enhance firmness in the submental region or along the nasolabial folds. Hence, the idea of combining ART with the use of special energy-based devices specifically designed to act regeneratively on the aged dermis, as well as on the lax fasciae and fibrous septa of the subcutaneous tissue. These electronic systems, not capable of increasing lost volumes on their own, find in ART a winning combination, precisely because together they act by recreating a more defined and toned

D. Piccolo et al., *A.R.T. Autologous Regenerative Therapy in Aesthetic Medicine*, https://doi.org/10.1007/978-3-032-05835-5_4

container, with renewed quality content and improved quantity with the restoration of lost volumes. The field of Energy-Based Devices (EBDs) is vast and continuously evolving; however, those relevant to the present context are devices designed to promote tissue regeneration by stimulating neocollagenesis within the dermis and subcutaneous tissue, where fibroblasts responsible for elastin and collagen synthesis are located. [4, 5] In this chapter, therefore, we will deal with those EBDs which, through a tissue heating mechanism, cause a controlled micro-damage, which will be repaired by the receiving organism through the production of new collagen. There are various operating principles of these devices (monopolar, bipolar transdermal, needle, surgical radiofrequencies, micro-focused ultrasounds, lasers, and others), and although it is impossible to review all of them exhaustively, we will try to precisely identify their mechanisms of action, the differences that characterize them and their execution techniques, so that the reader has the opportunity to fully evaluate how to combine them with ART. We will also address the correct selection of patients and possible complications, as well as the treatment of the latter.

4.2 Introduction

Whenever doctors are about to use a new EBD, they must not forget that, even before examining the physical characteristics of the device as well as its protocols of use, some basic essential notions are to be considered in order to obtain good results, but above all to avoid possible complications [6]. Among these notions, in the first place we find an anatomical knowledge of the areas we are about to treat. Secondly, it is important to have a thorough knowledge of the physiology of the skin aging process and, therefore, how to carry out a correct selection of patients so that the established treatment plan is personalized and most suitable for the specific case. With regard to the anatomy of the face and neck, it is important to recall that the relevant structural layers include the epidermis, dermis, subcutaneous tissue, and underlying musculature (Fig. 4.1).

Our determined action will focus on these structures, guided by ART and the use of specific EBDs [7]. The epidermis may present classical signs of aging, including reduced elasticity, diminished radiance, and the appearance of wrinkles and previously absent skin folds. The dermis, by contrast, is characterized by a distinct orientation of its elastic fibers, contributing to its structural stability despite repeated contractions of the underlying musculature. Recognition of this fiber orientation is essential, as it informs the appropriate treatment parameters for the selected EBD. The orientation of these tension lines was originally described by Langer in the late 1800s (Fig. 4.2) and remains a critical anatomical reference in current clinical practice.

The subcutaneous tissue consists of fascial structures and fibrous septa that keep the dermis adhered to the underlying muscular structures. These fibrous structures, composed of collagen, also undergo the skin aging process, with a progressive loss of tone and a consequent downward sliding of tissues due to gravity, resulting in the

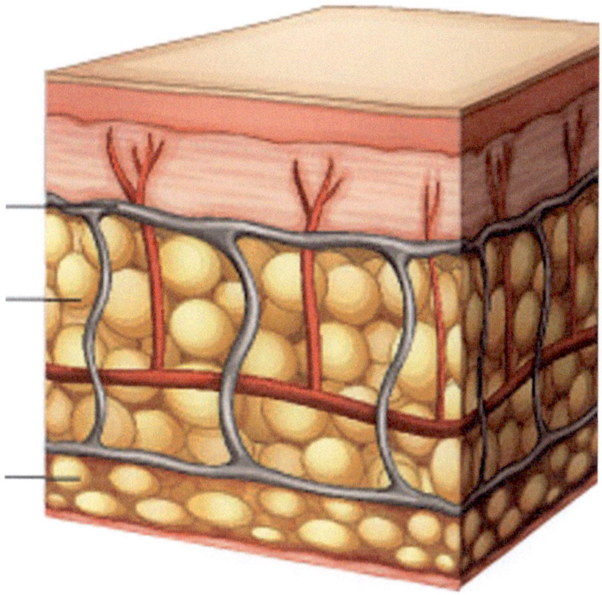

Fig. 4.1 Cross-sectional diagram illustrating epidermis, dermis, and subcutaneous layers

Fig. 4.2 Schematic showing Langer's lines and natural skin tension patterns

appearance of skin laxity. Over the years (Fig. 4.3), the underlying muscular structures also experience a reduction in volume and loss of tone, as do the underlying bone structures, which lose mass, accentuating the discrepancy between container and content.

In the pathophysiology of skin aging, the alteration of static and dynamic compartments of adipose tissue will find a place of primary importance, with some areas experiencing volume loss, such as the temporal, zygomatic, and malar areas, while

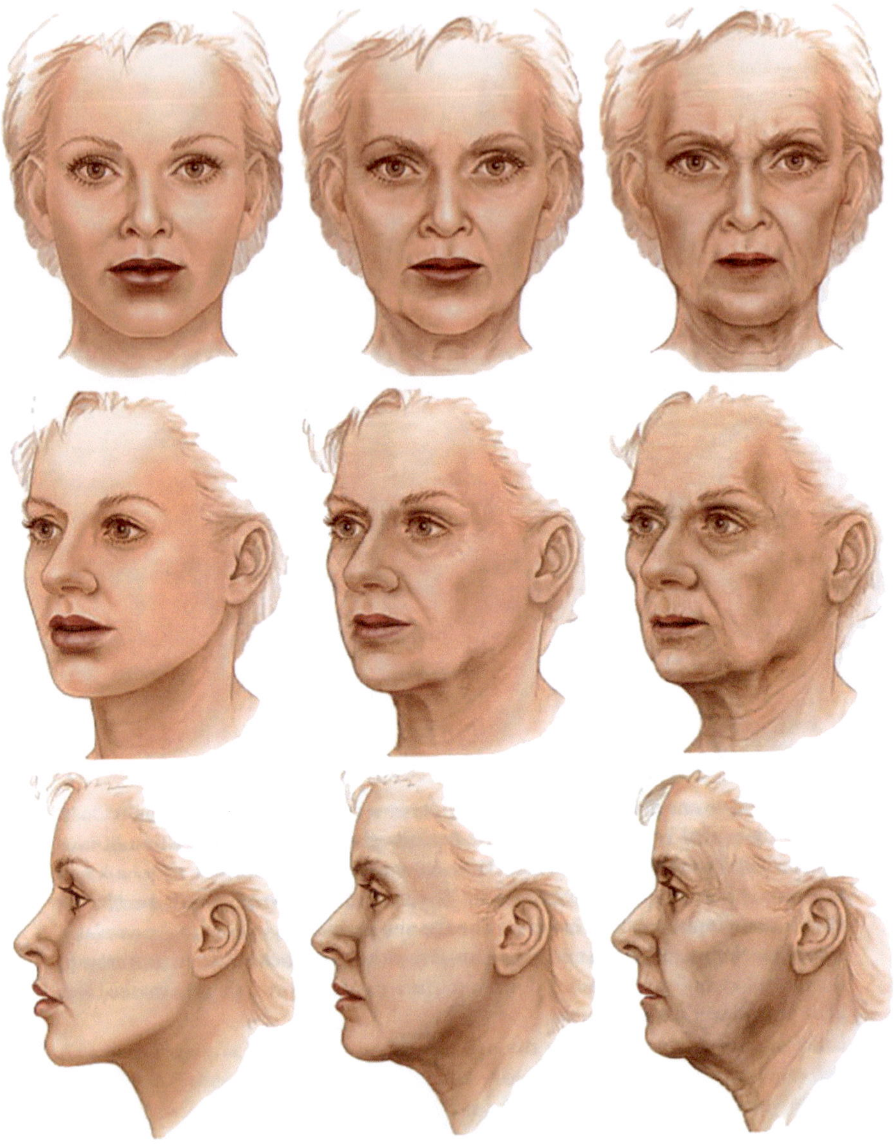

Fig. 4.3 Aging progression shown in frontal, three-quarter, and profile views

an increase in volume can be observed in nasolabial folds, jawline, and submental area (Fig. 4.4).

A careful objective examination of the patient, therefore, will be essential to understand in which areas to proceed with volumizing and/or regenerative treatments and, in which, instead, to act by restoring greater compactness and tension of the tissues through EBD. It should be remembered, however, that to trigger a correct

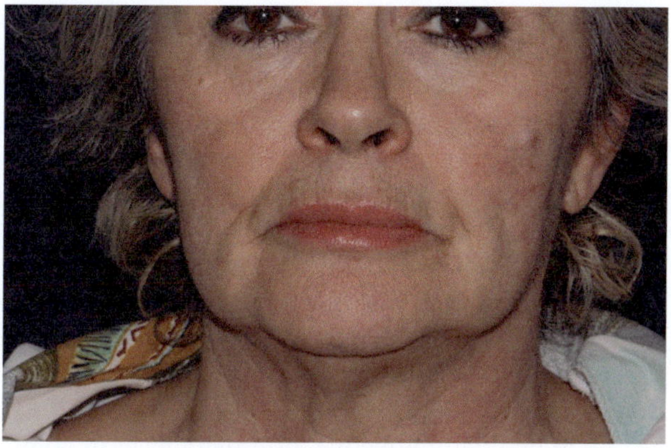

Fig. 4.4 Close-up highlighting wrinkles and sagging in aging facial skin

rejuvenation process of the face and neck, regenerative medicine is entirely based on neocollagenesis, which will involve both the epidermis and dermis on the surface, and the deeper subcutaneous tissue. Previously, we referred to tissue heating by means of an energy source (radiofrequency, ultrasound, laser) as the mechanism of action to trigger the neocollagenesis process. However, let us examine this process in more detail: it has now been scientifically established that to obtain a process of fibroblast stimulation, the triple-helix polypeptide chain of preexisting collagen has to be altered first. This occurs if a temperature between 50–55 °C and 65 °C is achieved in the target tissue for a brief and repeated time. These multiple thermal stresses end up damaging the structural characteristics of the tissue and inducing a reparative process with the production of new type III and I collagen, with reorganization of the fibers that will acquire the morphological appearance of young tissue not compromised by skin aging. The process is lengthy to unfold, usually between 6 and 8 months, and this clinically determines a progressive improvement in facial contours, thus counteracting natural and progressive skin aging. It is important to emphasize how the synergy between ART and EBD allows, today, to restore compactness and luminosity to the tissues, reduce the presence of superficial wrinkles, and restore lost volumes, without altering the patient's physiognomy, as required by the most modern aesthetic medicine. The type of EBD, therefore, that will be examined in this chapter concerns instruments capable of obtaining controlled tissue heating, equipped with safety systems for the patient and with a mechanism of action supported by validated scientific publications.

4.3 Capacitive Monopolar Radiofrequency and Autologous Regenerative Therapy

Initial clinical use of devices capable of promoting tissue regeneration through thermal mechanisms dates back to 2002, with the introduction of the first Capacitive Monopolar Radiofrequency system—originally launched in the United States the previous year. This technology was among the earliest capable of inducing neocollagenesis via controlled thermal stimulation and subsequent reparative biological responses. It rapidly gained widespread adoption and soon became a recognized brand in the aesthetic field. Its distinctive features included extensive scientific documentation and clinical studies, reflecting a strong emphasis on patient safety. The device applied radiofrequency energy to produce targeted heating, thereby initiating a regenerative process that yielded visible aesthetic improvements, while integrated control systems helped minimize the risk of thermal injury to the patient.

Following the introduction of this device, numerous radiofrequency modalities have been developed over the years—including monopolar, transcutaneous bipolar, subcutaneous, and needle-based systems. Among these, Capacitive Monopolar Radiofrequency has consistently demonstrated a favorable safety profile and has proven to be highly effective, offering a level of comfort during treatment that has led many patients to undergo repeated sessions over time. Furthermore, over time, the succession of four increasingly evolved generations of the device has led to progressive technical improvement, both by updating the algorithm that regulates energy emission to optimize results and by minimizing the discomfort generated by the treatment and the consequent downtime.

Setting aside the purely technical aspects of this EBD, as it is not the subject of this chapter, regarding its mechanism of action, we can say that it comes very close to that described for Micro-focused Ultrasounds.

In Capacitive Monopolar Radiofrequency, we do not have an incorporated ultrasonic tissue viewing system (which indeed makes HIFU—High-Intensity Focused Ultrasound unique in its kind), but in reality, the action of radiofrequency can be compared to a homogeneous thermal wave, which goes more or less deep depending on the tip selected on the handpiece (Fig. 4.5).

The RF reaches the deepest tissues, generating heating precisely where we find the highest concentration of collagen (dermis and fibrous septa present in the subcutaneous tissue) [8, 9], while on the surface, cooling generated by a cryogenic gas, accompanied by a vibration mechanism, makes the treatment more comfortable for the patient.

Moreover, the peculiar characteristic of this type of RF compared to others on the market is the maintenance of the heating wave for over a minute from the emission

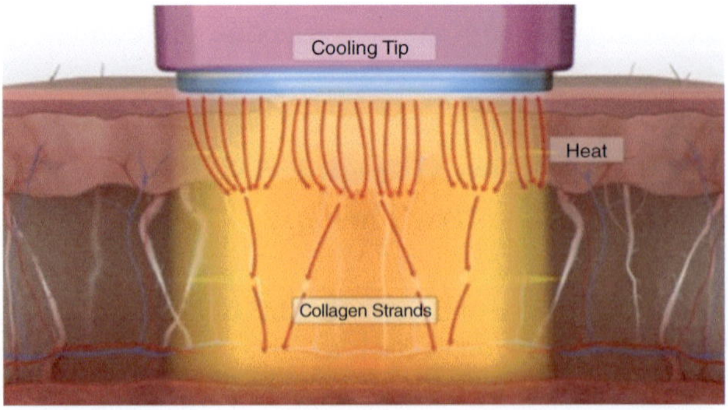

Fig. 4.5 Skin cross-section illustrating heat-based collagen stimulation with cooling tip

of a single spot. In this way, the denaturation of collagen protein chains is much more effective.

All this is carried out in complete safety, thanks to software control of the temperature on the handpiece tip and on the patient's epidermis.

Exceeding safety limits immediately generates an error signal from the device, inviting the operator to wait or move to another area, thus avoiding any possible thermal damage. Thanks to this control system, it is possible to carry out treatments at any time of the year, without risks related to sun exposure and on any photo-type, with virtually no downtime.

Studies with a thermographic camera have specified in detail the distribution of the generated heat: its distribution in the subcutaneous tissue is 1.5 to 3.5 times higher in the fibrous septa than in the adjacent adipose tissue, rich in water. This also explains why there is a three-dimensional contraction of the tissues and, consequently, an improvement in skin laxity.

Other scientific studies, then, have been directed at the histological behavior of tissues treated with RF (Fig. 4.6).

In one of these published studies, it is possible to observe how, after just a single treatment, there is a re-densification of collagen fibers that reaches its peak level after 6 months. Furthermore, histologically, we witness an increase in neocollagenesis in the months following the treatment. [10, 11].

It is precisely on these foundations that a radiofrequency treatment for specific body areas has been developed, where tissue tone loss and thinning prevail, with a predominance of lack of elasticity. This process, which may be a consequence of skin aging, but also due to significant weight loss or the aftermath of pregnancy, can be excellently addressed, after proper patient selection, in areas such as the inner arms, knees, and peri-umbilical region, for example. For these areas, special tips have been created that concentrate thermal energy precisely in the subcutaneous tissue, as well as more superficially, so as to trigger the production of new collagen capable of improving skin appearance and reducing skin laxity. All this in a single session, the effects of which will be enhanced in the subsequent 6–8 months, with maintenance of the results obtained for about 12–18 months (Fig. 4.7).

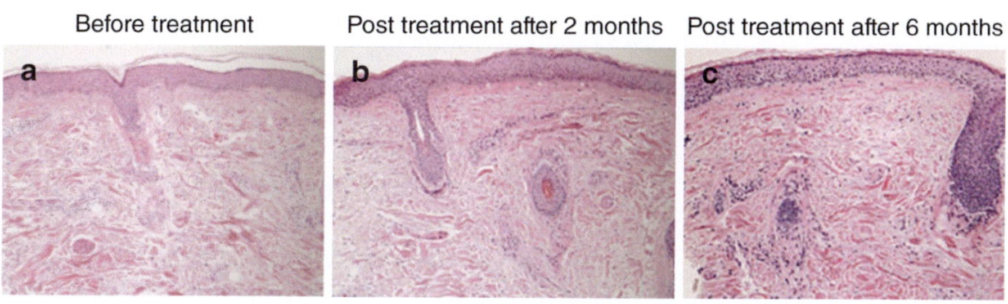

Fig. 4.6 Histological sections of skin tissue illustrating morphological changes over time following cosmetic dermatology treatment. (**a**) Baseline skin architecture prior to treatment. (**b**) Posttreatment at 2 months, showing mild improvement in dermal structure. (**c**) Posttreatment at 6 months, demonstrating increased collagen density and enhanced overall skin architecture

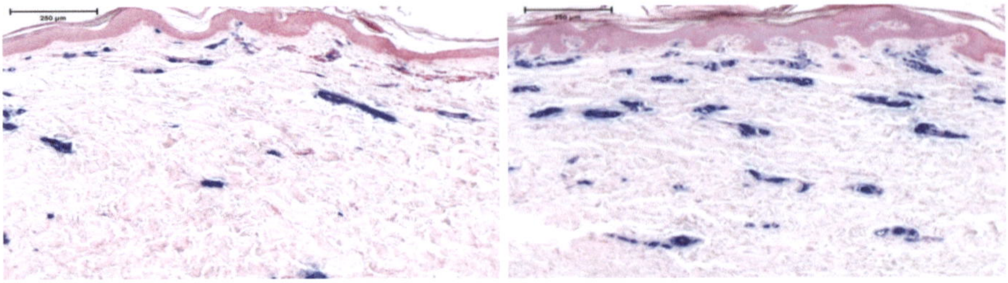

Fig. 4.7 Histological skin comparison showing structural differences in dermal staining

There is, then, another important difference in execution between Capacitive Monopolar Radiofrequency and HIFU, which deserves to be addressed. Since Capacitive Monopolar Radiofrequency emits a homogeneous thermal wave in the affected area, and not infinite coagulation points, as ultrasound does, it is possible to perform the treatment by delivering radiofrequency while the handpiece is moving on the skin surface. This results in dynamic heating that is evenly distributed over the entire chosen skin region, even when delivering very high energies, but without causing risks of burns.

At this point, it is important to understand which vectors to follow to obtain an optimal result, and here the studies carried out by Karl Langer, and made known in 1861 on the distribution of tension lines, come to our aid (Fig. 4.8).

Langer's lines correspond to the direction in which the natural tension vector of human skin is lowest. Through a meticulous histological study conducted across the entire skin surface, Langer was able to definitively identify these vectors for all areas of the face and body, thus creating a graphical representation that, even today, serves as an important guide for plastic surgeons, dermatologists, and aesthetic physicians.

Cox (1941) and Stark (1977) both confirmed that the direction of Langer's lines, revealed through an extremely high number of skin punch biopsies on cadavers, remained correct even after the skin was removed from the body, thereby demonstrating that the lines have an anatomical basis (Fig. 4.9).

Based on these considerations, it is possible to deliver our radiofrequency treatment by keeping our handpiece in motion, following the vectors of Langer's lines. This will create a more uniform stimulus for neocollagenesis, parallel to the direction of the elastic collagen fibers.

This technique proves particularly suitable when treating very thin tissues with little subcutaneous fat, such as the inner region of the arms or the knees (Fig. 4.10).

This is an example of a monopolar radiofrequency treatment followed, after about 30 days, by a treatment with Autologous Regenerative Therapy (about 12 cc per side in the zygomatic and malar area) (Figs. 4.11, 4.12, 4.13, and 4.14).

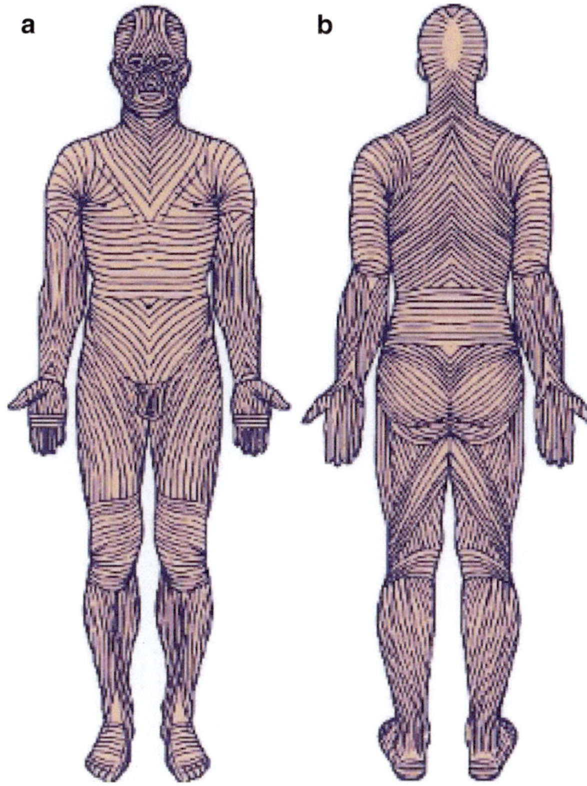

Fig. 4.8 Illustration of human muscle anatomy showing two views: the front (**a**) and back (**b**) of the body. The image highlights muscle fiber directions with detailed line patterns across the entire body, including the head, torso, arms, and legs

4.3.1 Microfocused Ultrasound

The use of ultrasound in the medical field began around the 1940s, primarily for diagnostic purposes through ultrasonography. Subsequently, this sector has continuously expanded and improved its reliability, thanks to technological developments that have achieved previously inconceivable image sensitivity. Later, ultrasound was applied to other areas of medicine, such as the treatment of certain neoplasms. Researchers' attention then turned to the application of ultrasound in aesthetic medicine and plastic surgery [12–14].

In the early 2000s, the first High Intensity Focused (or Microfocused) Ultrasound devices, commonly known as HIFU, were developed for nonsurgical treatment to correct signs of skin aging. In fact, around 2004, a group of American researchers demonstrated how Microfocused Ultrasound concentrated on the dermis and subcutaneous tissue raises their temperature, stimulating the production of collagen and elastin, essential elements for maintaining tissue tone and elasticity (Fig. 4.15) The peculiarity that differentiates HIFU devices from other machines that exploit the temperature mechanism is that the former achieve a concentration of tissue heating

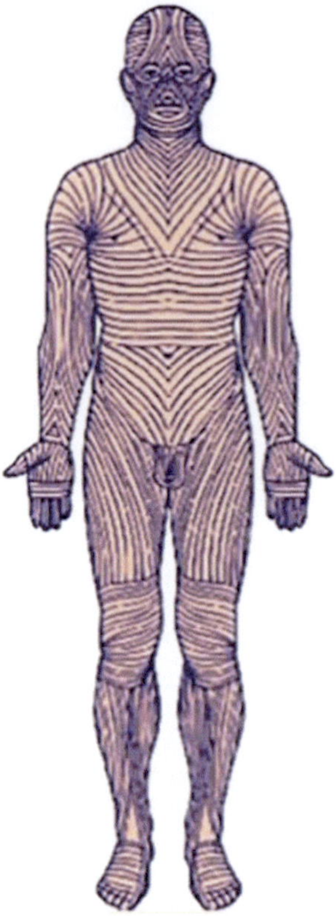

Fig. 4.9 Front view showing muscle fiber structure and direction

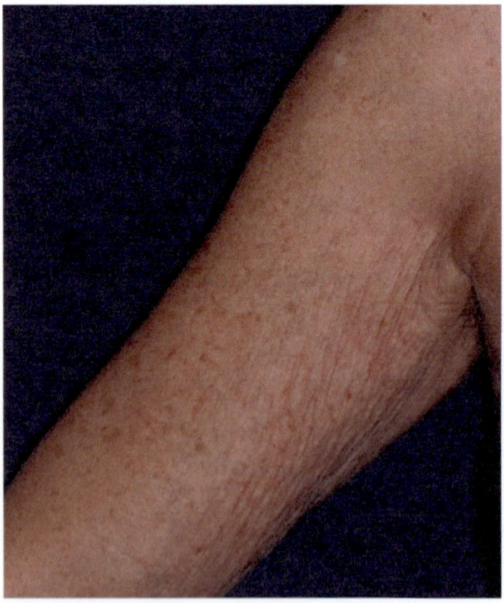

Fig. 4.10 Close-up of aged arm with wrinkles and age spots visible

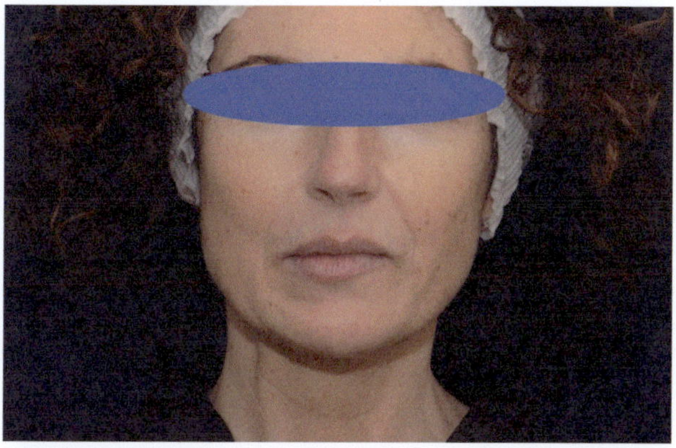

Fig. 4.11 Pretreatment front view

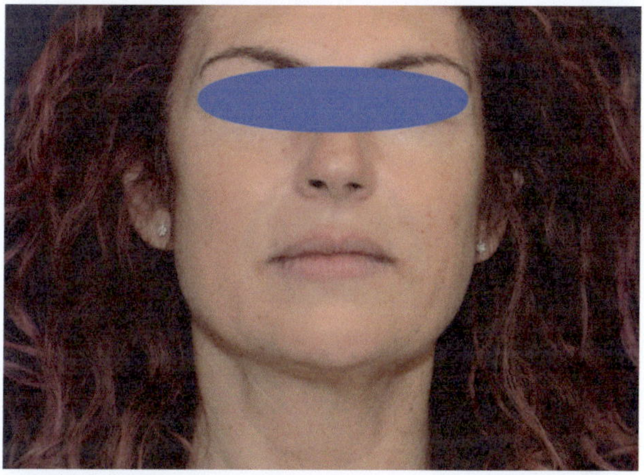

Fig. 4.12 Posttreatment front view

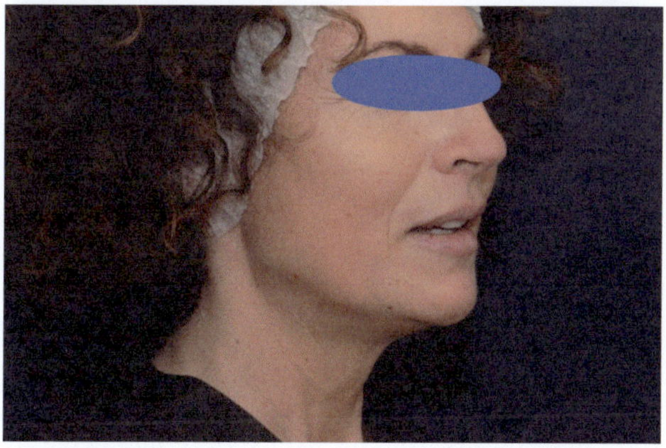

Fig. 4.13 Pretreatment side view

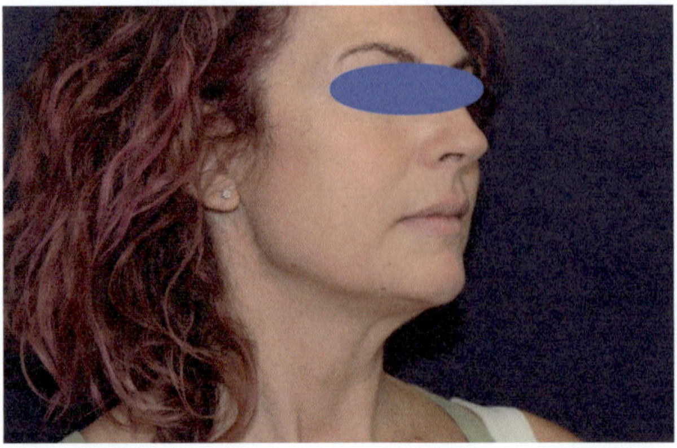

Fig. 4.14 Posttreatment side view

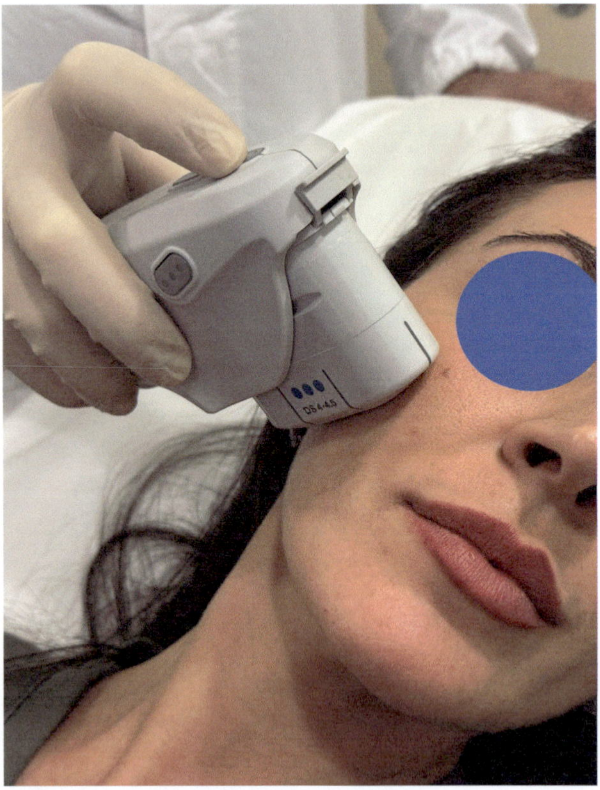

Fig. 4.15 Cosmetic procedure using microfocused ultrasound device on cheek in sterile setting

at a precise point, while with others, such as radiofrequencies, there is a thermal wave that spreads from the skin surface into depth, progressively dispersing as it descends into the tissues.

This image (Fig. 4.16) depicts an experimental model in which ultrasound waves, emitted by a specific transducer, are focused at a depth of 1.8 mm within a water

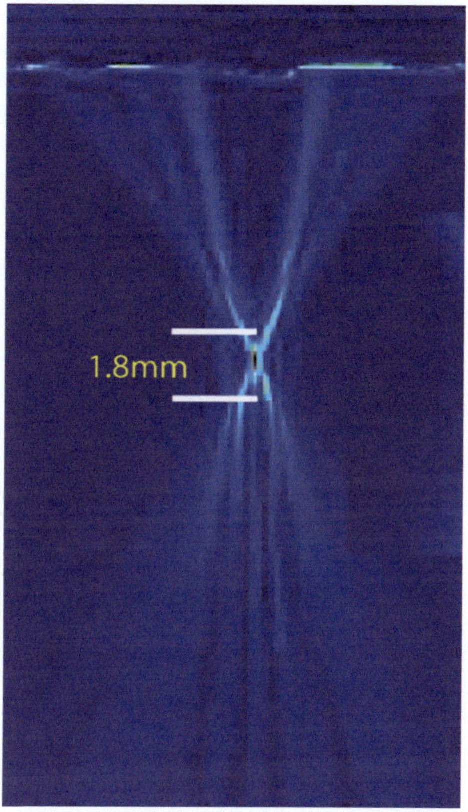

Fig. 4.16 Imaging output showing 1.8 mm focal point with energy convergence

medium. Using a specialized imaging system, the ultrasound energy is shown to concentrate 95% of its focus at an extremely precise point, while diverging immediately above and below the targeted depth, thereby losing its thermal effect. This enables highly accurate determination of the intended treatment depth. The heating effect is thus confined to a very specific and brief focal zone, addressing concerns regarding unwanted spread of thermal energy to adjacent sensitive structures. This instrumental approach provides clinicians with a high degree of procedural control.

The temperature induced by focused ultrasound typically ranges between 60 °C and 70 °C, creating thermal coagulation zones that are recognized by the body as injury and subsequently repaired through collagen production. This initiates a progressive and long-lasting skin rejuvenation process. These findings led to the development of various transducers designed to act at multiple tissue depths, targeting structures involved in collagen and elastin regeneration. The resulting controlled injury stimulates the production of new collagen and elastic fibers while preserving surrounding healthy tissues. Research in collagen biophysics has confirmed that at temperatures between 55 °C and 70 °C, the collagen helix undergoes contraction and denaturation [15–17], thereby initiating the neocollagenesis process [18, 19]. In contrast, other EBDs may not consistently achieve optimal thermal thresholds;

insufficient temperatures fail to initiate neocollagenesis, whereas excessive heat may result in significant or even irreversible tissue damage.

In this image (Fig. 4.17), a thermal coagulation zone is shown at considerable magnification, consisting of Thermal Coagulation Points alternating with areas of healthy tissue. The Coagulation Point will be repaired through the regeneration of collagen, which proceeds from nearby areas of healthy tissue. This process confirms the necessity of maintaining intact healthy interval areas; otherwise, the regenerative process cannot be effective.

Depending on the transducer model used, we will have Coagulation Points of various diameters, interspersed with different spaces of healthy tissue: the larger the diameter of the Coagulation Point, the greater the space of healthy tissue between one and another [20, 21].

However, the natural aging process of tissues does not only involve the elastic and collagen fibers of the dermis but also concerns the subcutaneous tissue, which undergoes a loss of tone in the fascia and fibrous septa, also composed of collagen and elastin, which maintain its tone. This results in the sliding of skin and subcutaneous tissue over the deep planes, with a loss of definition in facial contours.

Therefore, Microfocused Ultrasound has been designed to act with different transducers at various depths in the tissues, so as to trigger the regenerative process on multiple levels. An action aimed exclusively at heating only superficial tissues would have only partially resolved a much more complex process such as aging.

However, since there are considerable variations in the thickness of facial tissues, based on sex (male or female), race, or BMI, and even in the same individual there are differences in the location of anatomical structures between one area and another, it has become essential to "see" these tissues by always resorting to ultrasound-based imaging.

Thus, in 2009, the HIFU system was born, obtaining recognition from the American FDA for eyebrow lifting in the same year. This was followed in 2012 by recognition for lifting of the submental area and neck, in 2013 for ultrasound

Fig. 4.17 Microscopic tissue image showing linear bright points on irregular surface

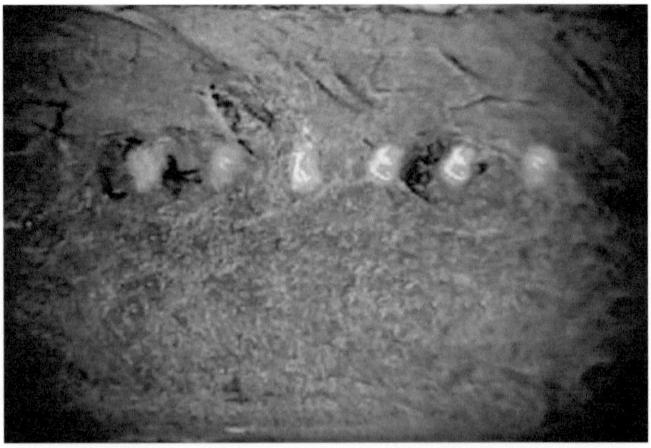

visualization of tissues ensuring correct depth of treatment, and in 2014 for the improvement of wrinkles and décolletage.

HIFU is an energy device that manages ultrasound in a first phase to study our patient's tissues, so as to program the treatment to be performed, and in a second phase, by changing the delivery mode to microfocused, to stimulate our tissues through the heating process.

Unique in its kind, HIFU is equipped with a viewing system called Deep See, which has obtained recognition from the American FDA, capable of defining the boundaries between dermis, subcutaneous tissue, muscle fascia, as well as highlighting vessels or bone structures, and injectable material implants that should be avoided during treatment.

Furthermore, the Deep See system offers the significant advantage of providing real-time visual feedback, enabling verification that the Microfocused Ultrasound generator is being applied correctly and that the treatment is being performed under optimal conditions.

In order to treat tissues at multiple depth levels, three transducers have been developed that act at 1.5 mm, 3.0 mm, and 4.5 mm, respectively. The first, obviously, is more superficial and dedicated to stimulating the dermis with a direct action on the production of new collagen and elastin. The second, at 3 mm depth, acts on the subcutaneous tissue and particularly on fascia and septa present here, as well as on muscle fascia in those anatomical areas that have a reduced thickness. Finally, the 4.5 mm depth transducer delivers its energy as many points of contraction of the muscle fascia, to restore better tone and consistency (Fig. 4.18).

A more in-depth analysis of this topic identifies the Superficial Musculoaponeurotic System (SMAS) as a fundamentally important target in the treatment of facial skin aging.

For the neck, on the other hand, it is the platysma muscle that proves responsible for the sagging of tissues, especially in the submental region (Fig. 4.19). At this point, the ultrasound imaging system becomes essential to verify the position of the muscular fascia of these structures, where our ultrasound action will focus, in order to achieve its structural reconsolidation through Thermal Coagulation Points (Fig. 4.20).

The notable anatomical differences in these areas, which vary from subject to subject, make it very difficult to standardize the execution technique. Rather, they require personalization, which can only be achieved through a preliminary ultrasound assessment.

The HIFU system was also launched on the global market with a well-defined treatment protocol, which indicated the areas to be treated and the number of lines (consisting of the set of points they contain) to be delivered for each traced area, varying according to the type of transducer used (Fig. 4.21).

Initially, HIFU was promoted as the EBD capable of achieving results similar to a facelift, but without surgery. In reality, with subsequent scientific insights regarding regenerative medicine [22–24], which today increasingly plays an essential role in correcting skin aging, it has been understood that this treatment is not only incomparable to a facelift but has a remarkable regenerative power on target tissues, especially when combined with Autologous Regenerative Therapy.

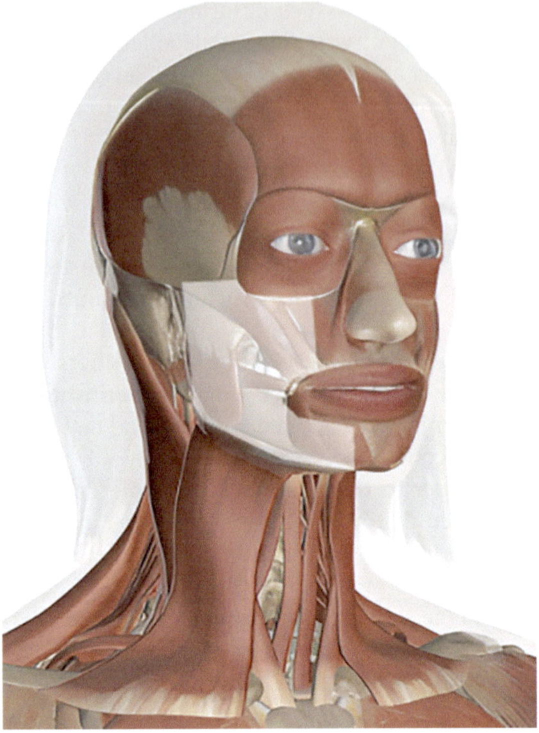

Fig. 4.18 Anatomical illustration showing facial and neck muscles beneath transparent skin

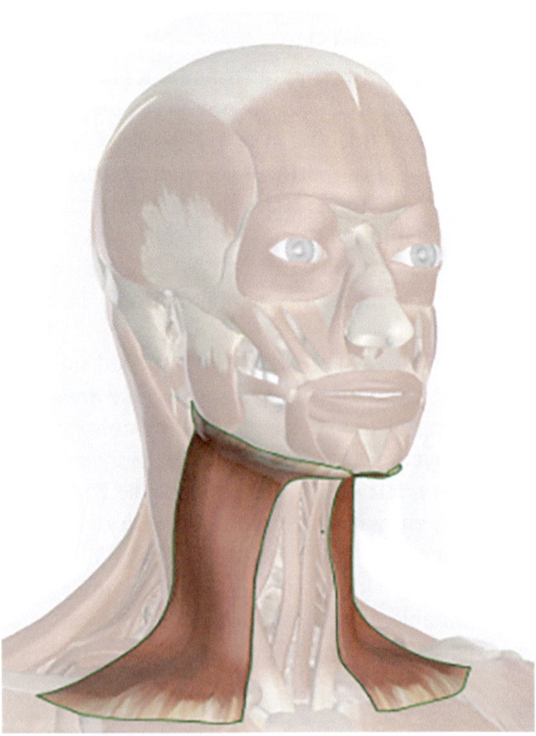

Fig. 4.19 Anatomical view highlighting platysma muscle with underlying structures visible

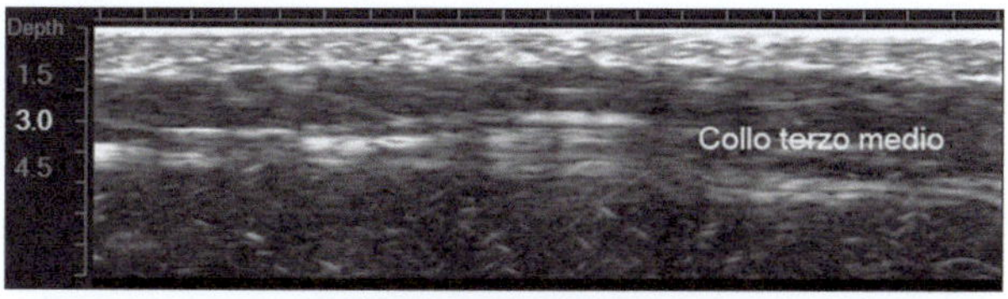

Fig. 4.20 Ultrasound of mid-neck showing tissue cross-section

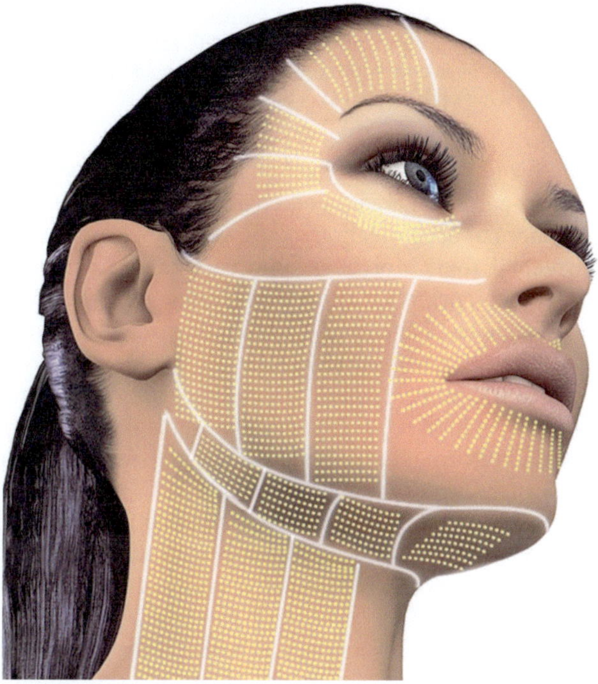

Fig. 4.21 Illustration showing treatment zones on face and neck with dotted patterns

After gaining considerable experience with HIFU treatments not associated with anything else, I have found that with good patient selection and proper application of the treatment protocol, enthusiastic results can be obtained in tissue recompaction. The treatment involves a single session, generally lasting less than an hour, with progressive improvement of tissues for at least 6 months and maintenance of the result obtained for 12–18 months (Fig. 4.22).

This treatment, therefore, represents a valid alternative for those who still have initial forms of skin laxity or do not yet feel ready for a major surgical procedure such as a facelift. It can also be performed in the height of summer and does not generate any convalescence.

But during patient selection, an important fact emerged: in many patients, especially those over 50–55 years of age, there is a gradual process of skin aging that involves all the anatomical structures of the face. Thus, alongside a loss of tissue tone, there is a depletion of certain adipose compartments, both static (deep) and dynamic (more superficial), resulting in the loss of volume in areas such as the

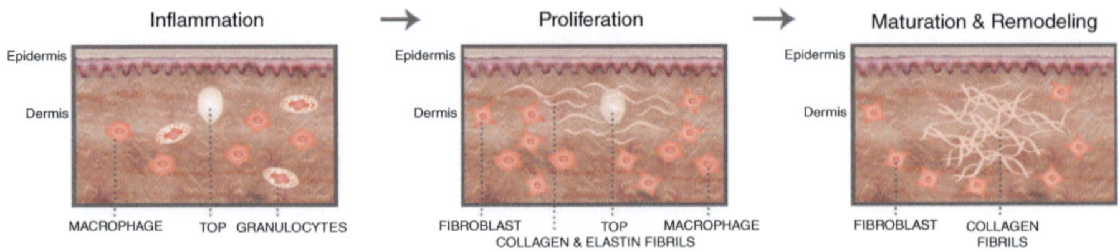

Fig. 4.22 The process of collagen formation and the role of different cells and proteins in skin structure. Inflammation: macrophages, TGF, and gamma-delta T cells. Proliferation: fibroblasts, TGF, macrophages, and collagen 6. Maturation and Remodeling: fibroblasts and collagen fibers, depicting a network formation

eyebrow region, the temporal region, the entire zygomatic-malar surface, as well as the peri-palpebral area. In other areas, however, we see a heaviness of tissues due to an abnormal accumulation of adipose tissue, such as in the mandibular area and the nasolabial fold. Moreover, with careful observation of the skin surface, we can detect the appearance of widespread fine wrinkles in the peri-palpebral, malar, and peri-oral areas, for example. The quality of the skin appears compromised, with loss of luminosity and elasticity.

Such a complex picture, therefore, requires articulated treatments that are certainly capable of repositioning tissues that have slipped downwards, while simultaneously restoring lost volume and deeply stimulating the tissues, so that the overall response becomes optimal.

At this stage, clinical considerations indicated the necessity of combining Autologous Regenerative Therapy with the effects of Microfocused Ultrasound. The initial approach focused on restoring the structural tone of the surrounding soft tissue envelope ("the container") by stimulating the production of elastic fibers and collagen, specifically in regions where these components had lost consistency and compactness. With the aid of ultrasound imaging, the therapeutic action of Microfocused Ultrasound was directed to the zygomatic region, the mandibular contour, and the submental and cervical areas.

This treatment was subsequently combined with the implantation of mesenchymal cells and growth factors into both superficial and deep facial layers. A more fluidized component was administered in areas requiring enhanced regenerative activity.

In anatomical regions with notable adipose volume depletion—such as the temporal or zygomatic-malar areas—treatment protocols included the use of a graft with a higher adipose concentration to provide additional volumization during the same session.

Due to the intrinsic fragility of stem cells during both the harvesting and implantation phases, a standardized technique was adopted. This approach is based on a rigorously defined protocol supported by numerous peer-reviewed scientific publications validating its clinical efficacy.

But beyond this, it was important to establish times and methods for the correct integration of the two methodologies.

First of all, it should be considered that both the technique using Microfocused Ultrasound and ART, since they act primarily on tissue regeneration, provide a response with a latency time of at least 45–60 days from treatment, then proceeding with the progressive improvement of tissues up to 6–8 months afterwards. For this reason, it becomes important to establish a timing for the integration of the two treatments that is not excessively extended over time, in order to concentrate their effects and, consequently, the final results.

Now, considering that HIFU causes a significant tissue inflammatory process in the 30–40 days after treatment, I preferred to schedule the ART implant after at least 30 days. This is to prevent the inflammatory cascade from compromising the vitality of the implanted cells.

EBD HIFU involves a single session, lasting about an hour and, after approximately 1 month, we will perform ART with a single implant using a microcannula on all the planned areas. Accurate photographic documentation will be crucial, which should be carried out on day 0 and at 90 and 180 days. The same program can then be repeated after 12–18 months (Figs. 4.23, 4.24, 4.25, 4.26, 4.27, and 4.28).

The rationale for integrating Autologous Regenerative Therapy (ART) with Energy-Based Devices (EBDs) warrants careful consideration. In clinical practice, numerous cases present with advanced stages of skin aging, particularly in areas such as the arms, knees, and inner thighs, where significant subcutaneous tissue depletion and diminished skin elasticity are observed. In such challenging

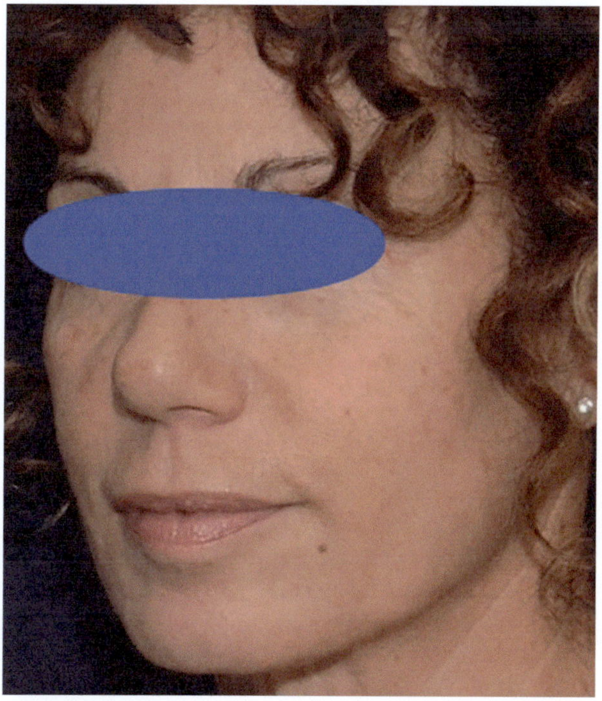

Fig. 4.23 Pretreatment side view

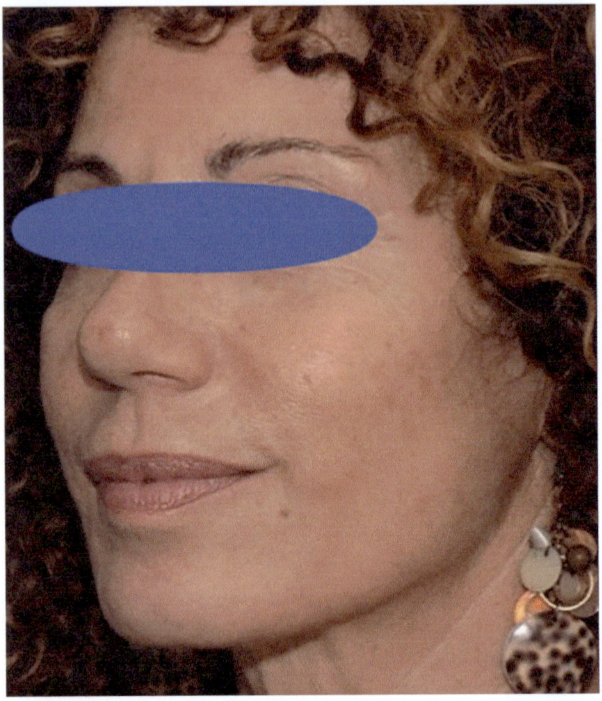

Fig. 4.24 Posttreatment side view

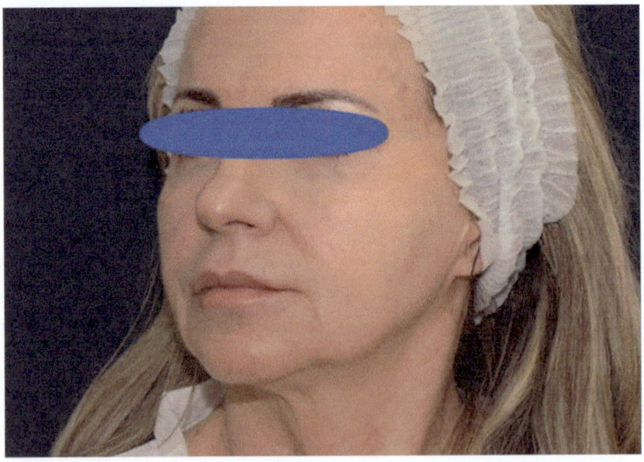

Fig. 4.25 Pretreatment side view

scenarios, treatment outcomes may appear limited, raising questions about therapeutic feasibility.

Nonetheless, by employing advanced technologies judiciously, it is possible to pursue treatment pathways that might otherwise be considered unviable. A detailed and objective patient assessment is essential in these instances. For example, if the hypodermal layer is severely depleted, it becomes unlikely that subcutaneous tissue compaction can be achieved using EBDs alone. Similarly, if the biological

Fig. 4.26 Posttreatment side view

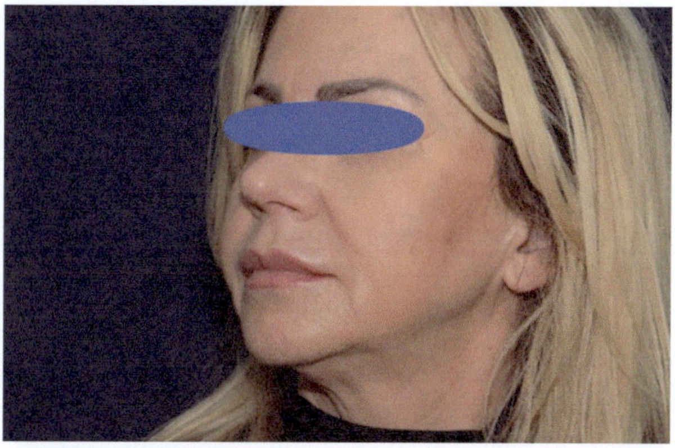

Fig. 4.27 Pretreatment front view

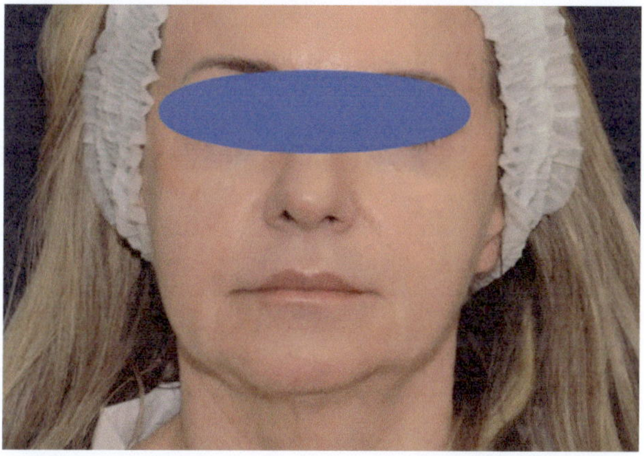

Fig. 4.28 Posttreatment front view

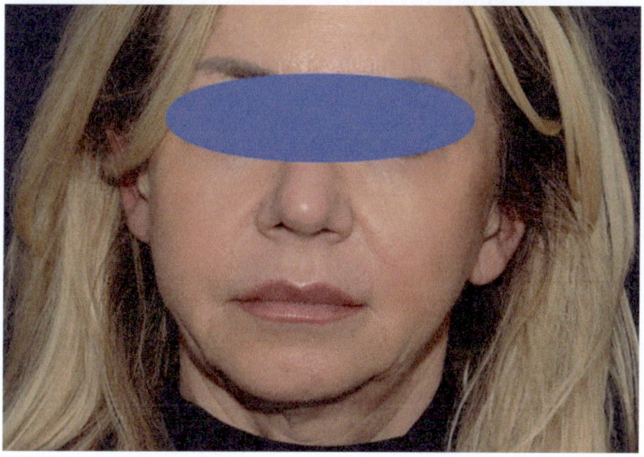

environment is compromised, the regenerative potential of mesenchymal cell implantation may be significantly reduced.

These limitations highlight the importance of a multimodal approach. The synergistic use of EBDs and ART offers a compelling strategy to address both structural and cellular deficiencies. ART allows for the selective enhancement of hypodermal volume through adipocyte-rich grafts or, alternatively, the stimulation of dermal regeneration in areas primarily affected by collagen and elastin loss. EBDs complement this action by inducing collagen production and remodeling in targeted regions, leading to tissue compaction and improved elasticity.

This integrated approach is also applicable to skin aging of the face, neck, décolleté, and hands. A thorough analysis of facial contours and volume distribution should always precede treatment, allowing for the formulation of a comprehensive procedural plan. This plan should specify the areas to be treated with EBDs, those suitable for regenerative cell therapy, and those requiring volumetric enhancement to counteract hypodermal hypotrophy associated with chronological aging.

Particular attention should be given to patients exhibiting severe photoaging, such as long-term smokers or individuals with excessive sun exposure, where dermal damage is often extensive. In such cases, the most favorable outcomes have been associated with the combined application of validated EBDs and ART, delivered using a standardized and reproducible protocol.

When combining these modalities, treatment sequencing is critical. Monopolar radiofrequency is typically applied to the entire face and neck in a single session. In contrast, high-intensity focused ultrasound (HIFU) may be selectively applied to the lower third of the face and neck, while autologous cell implantation can be simultaneously directed to the middle and upper thirds of the face.

This proves particularly useful when, above all, the patient needs a strong regenerative boost and, perhaps, also wishes to optimize response times.

In the context of Capacitive Monopolar Radiofrequency treatment, the protocol typically begins with the EBD session, which is known to induce a local inflammatory response. This is followed by fibroblast stimulation that promotes the synthesis of new collagen and elastin. After approximately 30–40 days—once the inflammatory phase has subsided—autologous cell implantation can be performed without compromising cell viability, thereby enhancing and accelerating the regenerative effect.

At this stage, it is also possible to evaluate the need for volumetric augmentation. The adipose material harvested for implantation may be processed with varying degrees of fluidity, depending on the anatomical area and the intended therapeutic outcome.

The evaluation of treatment results should not take place before 6 months have elapsed following the cellular implant. At that time, a decision can be made regarding the need for an additional ART session or a full repetition of the combined protocol, typically around 18 months after the initial radiofrequency treatment.

Patient engagement is essential in this process. Accurate photographic documentation facilitates the assessment of outcomes and supports shared decision-making regarding the potential for further interventions. It must be acknowledged that cutaneous aging is a continuous process; therefore, both clinicians and patients must determine whether and how proactively to address its progression.

4.4 Fiber Optic Laser 1470 NM and Target

The chapter author wishes to thank Monica Baldessin MD for her help and competence and sharing her clinical experience and photographs without which the chapter would not have been complete.

The laser (Fig. 4.30) is a diode device with an output power capacity ranging from 0.1 to 15 watts and a wavelength of 1470 nm. According to the theory of selective photothermolysis described by Anderson and Parish in 1983 [1], the 1470 nm wavelength specifically interacts with water and fat, sparing surrounding tissues from thermal damage [1] (Fig. 4.29).

Energy transmission can occur in pulsed or continuous mode depending on the pretreatment settings. The device is equipped with a software featuring various preset treatment programs. Several accessories are available, such as optical fibers of different calibers and a gynecological handpiece, making the equipment multifunctional. Its use has been validated in various minimally invasive surgical and nonsurgical procedures, particularly involving the integument, mucous membranes, and adipose tissue (aesthetic medicine and surgery, gynecology, general surgery, plastic surgery) [8, 9, 25, 26].

Sterile, single-use fibers with frontal or radial emission are used, with calibers ranging from 200 to 800 microns (Fig. 4.31).

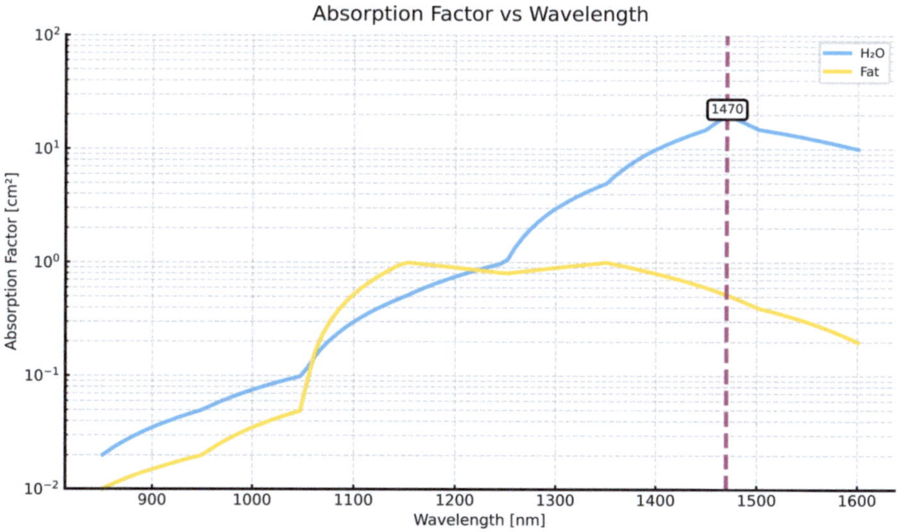

Fig. 4.29 Laser therapy, Vol 29(2)—June 2022

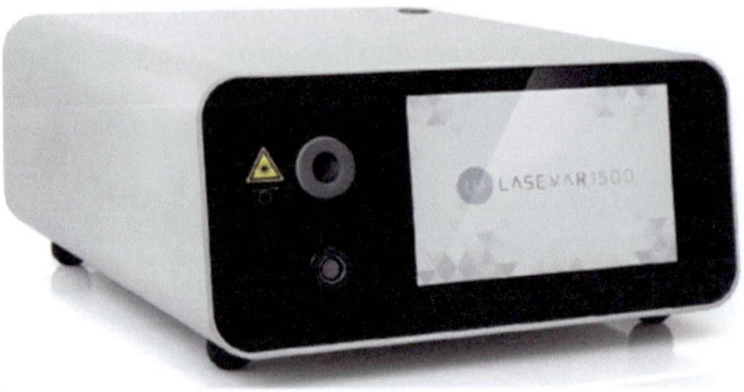

Fig. 4.30 The device is both CE marked and FDA approved

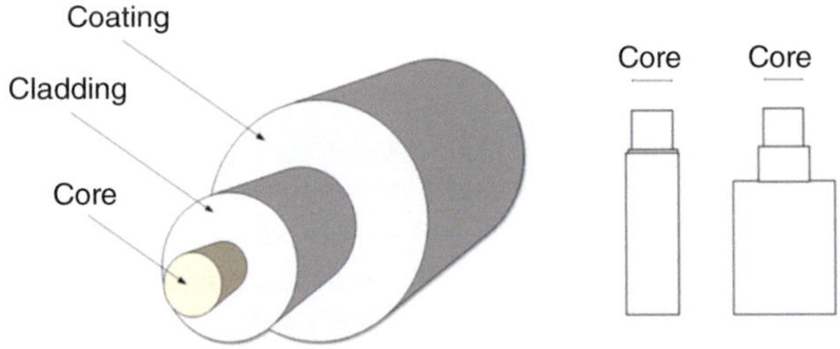

Fig. 4.31 The diameter of the optical fiber is expressed in microns and is represented by the central core through which the energy travels (core)

Frontal optical fibers emit energy from the anterior end at a 37-degree angle to the longitudinal axis, providing directionality and precision on target structures. They allow for the design of a hypodermic grid and the creation of retraction lines in areas of greater laxity.

Radial fibers emit energy at 360 degrees relative to the longitudinal axis, enabling the creation of a ring-shaped light pattern around the fiber. This design is particularly effective when a broader and more homogeneous distribution of thermal energy is required, such as in cases of diffuse cheek rugosity or double chin [11, 27] (Fig. 4.32).

The interstitial laser lifting is a mini-invasive procedure primarily used for treating laxity and skin aging. Thermal energy is delivered with extreme precision through a flexible optical fiber with a diameter of 200–300 microns inserted into the dermal layers or subcutaneous compartment. This causes an immediate contraction of the retinacula cutis [8], followed by stimulation of fibroblasts to produce new collagen fibers.

The creation of retraction lines and dermal remodeling improve the skin's adherence to the SMAS, resulting in more defined skin contours and visibly more homogeneous and compact skin [27–30] (Fig. 4.33).

This method is commonly used to treat various areas of the face, such as the periocular region, zygomatic-malar area, and lower third of the face, neck, and décolleté. Thanks to its tightening effect combined with lipolytic activity, it finds its maximum expression in remodeling the mandibular profile, jowl rugosity, and double chin [8, 10, 29–31].

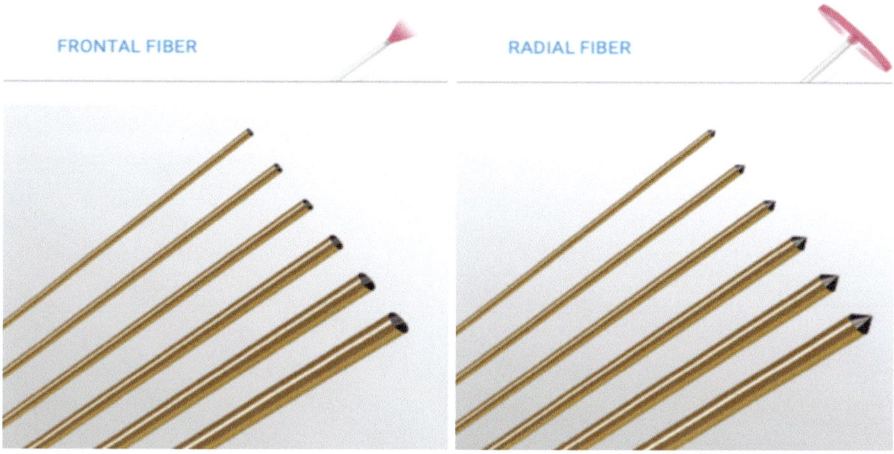

Fig. 4.32 Optical fibers available in two tip types and six different calibers

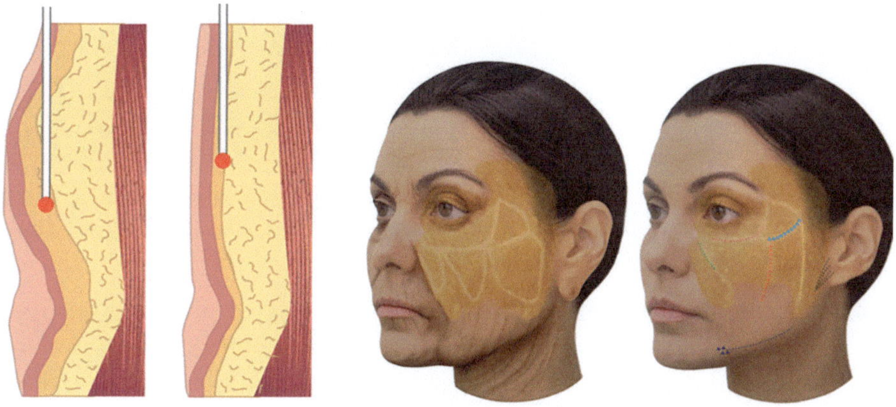

Fig. 4.33 Mechanism of action of the optical fiber in the SMAS

In lipolytic mode, using a radial emission optical fiber or one with a caliber greater than 400 microns, it allows for body contouring and treatment of localized adiposity [32–37].

The procedure is considered a nonsurgical lift as the patient can benefit from a minimally invasive outpatient procedure that does not interrupt work and social activities. The minimal downtime consists of modest edema and redness in the treated area, which resolves within a week. Transient bruising and dysesthesia may occasionally occur [8, 26, 32–34].

An example is presented of optical fiber laser treatment applied to the lower third of the face using a 300-micron optical fiber with frontal emission, combined with an Autologous Regenerative Therapy session after 30 days (9 cc per side in the temporal region, midface, and chin area) (Figs. 4.34, 4.35, and 4.36).

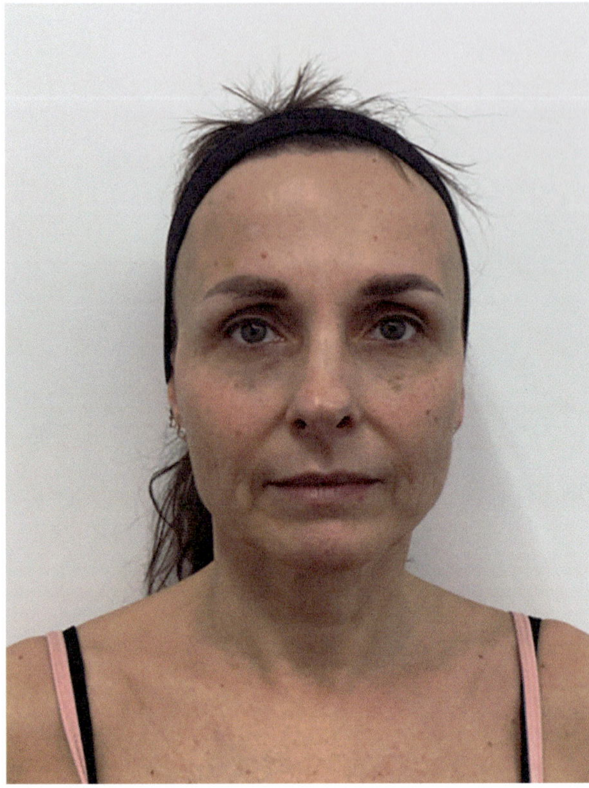

Fig. 4.34 Pretreatment front view

4.4.1 Fractional Laser

Non-ablative fractional laser is a laser treatment used to improve skin texture, reduce wrinkles, and stimulate collagen production.

Fractional lasers emit pulses of light energy that penetrate the dermal surface, overheating microscopic columns of tissue (Micro Thermal Zone MTZ) [38–41], comparable to pixels of a digital image (Fig. 4.37), while keeping the surrounding tissue intact. The controlled thermal damage triggers the natural wound healing response by fibroblasts, producing new collagen and elastin fibers. Collagen is a structural protein that, together with elastin and glycosaminoglycans, gives firmness and plumpness to the skin.

The lightscan handpiece with a wavelength of 1470 nm (Fig. 4.38) creates multiple MTZ overheating areas surrounded by areas of undamaged tissue. At the epidermal level, a micro-exfoliation occurs, resulting in significant cell renewal. At the dermal level, after an acute and transient phase of a few hours characterized by slight redness, a phase of remodeling of the skin texture follows.

Following resolution of the exfoliative phase, increased brightness and improved uniformity of the complexion are typically observed.

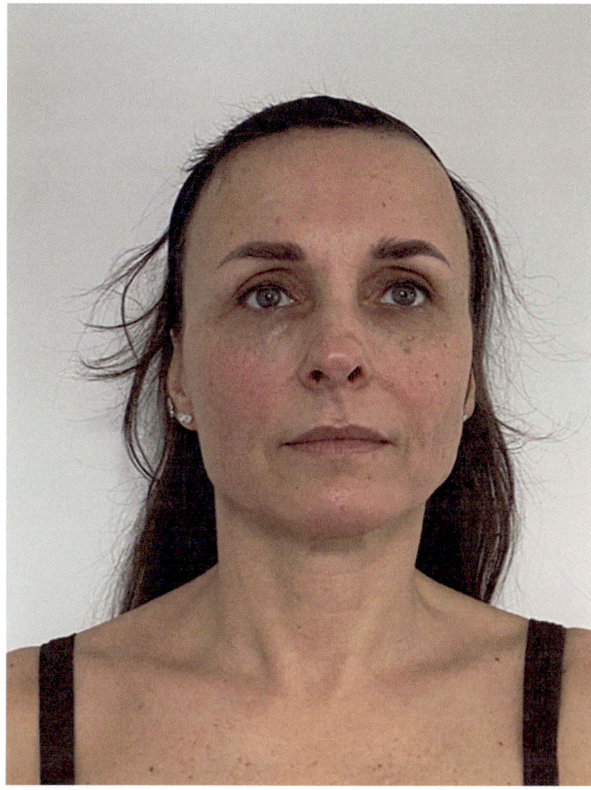

Fig. 4.35 Endolaser full face treatment, 300 micron fiber, 900 joules, 2.2 watts

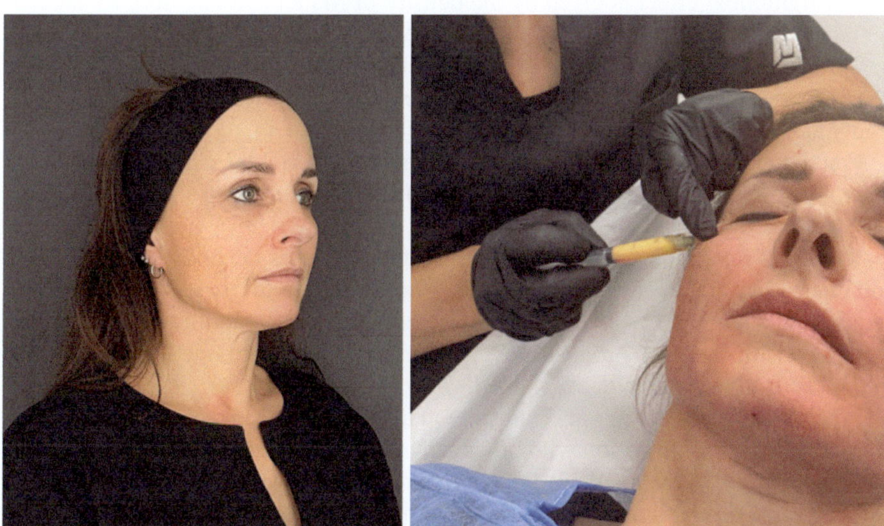

Fig. 4.36 Autologous Regenerative Therapy (ART) after optical fiber laser

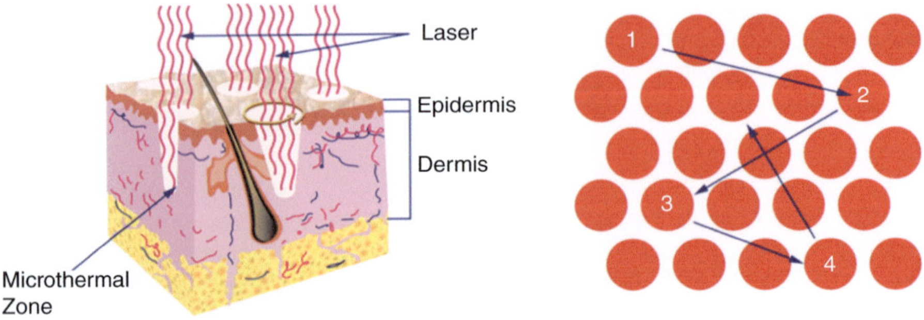

Fig. 4.37 Mechanism of action of non-ablative fractional laser on the skin

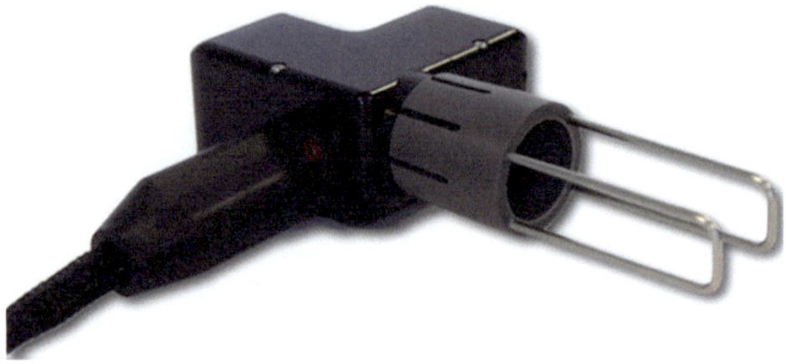

Fig. 4.38 Lightscan handpiece 1470 nm

Dermal redensification will be visible at the peak expression of neocollagenesis, which corresponds to the 28-day cell cycle.

The tightening effect is given by the density of MTZs and therefore by the percentage of treated skin (cumulative effect of MTZ action).

Three sessions are recommended at 3-week intervals.

Over time, dermal redensification is achieved, along with improved skin luminosity and texture, resulting in overall skin rejuvenation.

Fractional laser resurfacing can be used to treat fine wrinkles, acne scars, enlarged pores, skin discoloration, and imperfections. It is commonly performed on the face, neck, décolletage, and hands [26, 32–38].

Recovery time after fractional laser resurfacing varies depending on the depth and intensity of the treatment. Patients may experience mild redness in the immediate aftermath, sometimes associated with swelling, which generally resolves within a week. Results become progressively evident as collagen remodeling continues in the weeks following treatment.

Fractional laser resurfacing is considered a safe and effective procedure and can be performed in combination with other aesthetic medical treatments.

Figure 4.39 depicts the main types of skin damage created by the action of an ablative laser, non-ablative fractional laser, and ablative fractional laser.

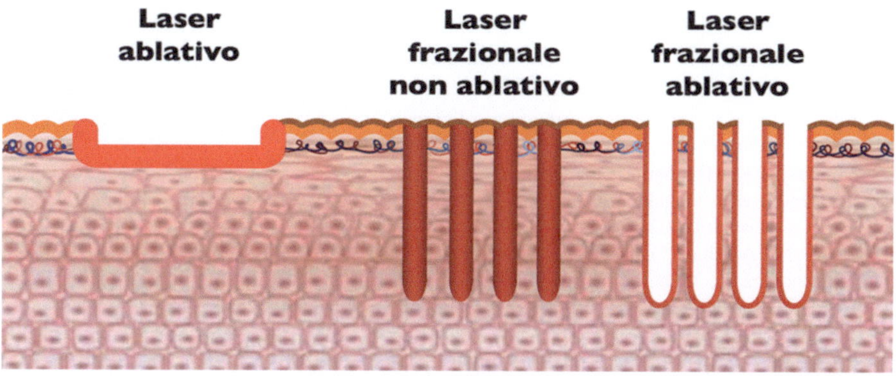

Fig. 4.39 Laser non-ablative fractional ablative fractional laser ##laser Ablative fractional laser

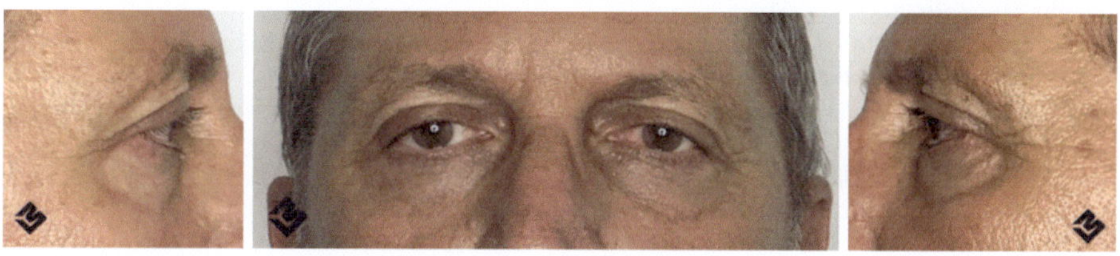

Fig. 4.40 Pretreatment right, front, and left view

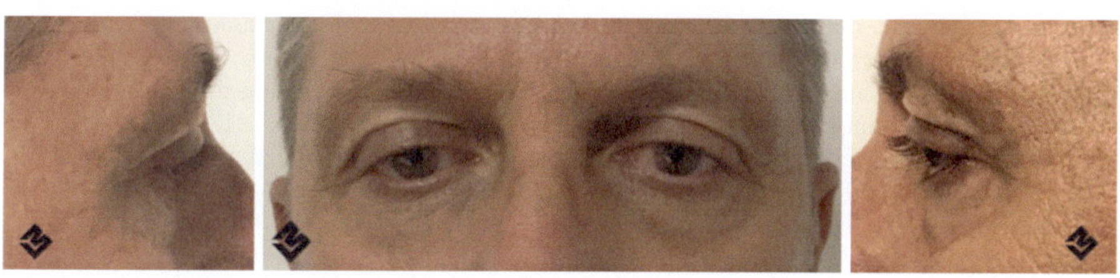

Fig. 4.41 Posttreatment right, front, and left view

The action of the ablative laser is limited to the superficial dermal layers, which are literally coagulated.

The non-ablative fractional laser, while exerting its action both superficially and in depth, allows for the preservation of almost all of the epidermis.

Ablative fractional laser exerts a superficial and deep parcelled ablation of the dermal layers [38–41].

Clinical Case: This is an example of non-ablative fractional laser treatment performed in the periocular region (three sessions at 3-week intervals). After about 30 days, the patient underwent a session of Autologous Regenerative Therapy with injection of about 5 cc per side (Figs. 4.40 and 4.41).

References

1. Anderson RR, Parish JA. Selective Photothermolysis: precise microsurgery by selective absorption of pulse radiation. Science. 1983;20:524. https://doi.org/10.1126/science.6836297.
2. Marquardt K, Hartmann C, Wegener F, Park JY, Halbert D, Hsu S, Hengl T. Microfocused ultrasound with visualization induces remodeling of collagen and elastin within the skin. J Cosmet Dermatol. 2024;24
3. Amiri M, Ajasllari G, Llane A, Casabona G, Pavicic T, Sevi J, Spada J, Vachiramon V, Vasconcelos R, Wah ST, Muka T, Fabi SG. Microfocused ultrasound with visualization (MFU-V) effectiveness and safety: a systematic review and meta-analysis. Aesthet Surg J. 2024;45:sjae228.
4. Wong A, Lowery AS, Bloom JD. Ultrasound therapy for the skin. Facial Plast Surg Clin North Am. 2023;31(4):503–10.
5. Al-Jumaily AM, Liaquat H, Paul S. Focused ultrasound for dermal applications. Ultrasound Med Biol. 2023;6:S0301-5629(23)00292-2. https://doi.org/10.1016/j.ultrasmedbio.2023.09.004.
6. Sanyal RD, Fabi SG. Energy-based devices for the treatment of facial skin conditions in skin of color. J Clin Aesthet Dermatol. 2024;17(6):22–32.
7. Chao JR, Porter JP, Hessler J. Cosmetic treatments with energy-based devices in skin of color. Facial Plast Surg. 2023;39(5):496–500.
8. Oliveira Modena DA, de Melo Yamamoto AP, Ferreira da Silva ETB. Endolift® is a non-surgical treatment for skin tissue conditions. Is there evidence for its application? Lasers Med Sci Rev. 2025;40(1):22.
9. Wassmer B, Zemmouri J, Rochon P, Mordon S. Comparative study of wavelengths for laser lipolysis. Photomed Laser Surg. 2010;28(2):185–8.
10. Wu X, Huang Q, Cai H, Zhang Y, Wang Z, Ma Z. Temperature-dependent variations in the proliferation rate, cell cycle distribution and expression of cell cycle-associated proteins in cultured human dermal fibroblasts. Arch Dermatol Res. n.d.;
11. Chen S, Pislaru C, Kinnick RR, Morrow DA, Kaufman KR. Evaluating the dynamic performance of a fibre optic pressure microsensor. Physiol Meas. 2005;26(4):N13–9.
12. Ramirez S, Scherz G, Leprince R. Pilot study: single-depth superficial ultrasound with subdermal injections of diluted calcium Hydroxylapatite for improving lower face skin quality. Plast Reconstr Surg Glob Open. 2024;12(10):e6210.
13. Bukhari A, Shadid A, Al-Omair A, Alheggi A, Almukhadeb EA, Albarqawi SY. Patient satisfaction following treatment with micro-focused ultrasound with visualization: a retrospective cross-sectional study. Skin Res Technol. 2024;30(8):e13917.
14. Pan R, Gu D, Zhang Y, Meng X, Liu T, Zhong H, Xu Y. Efficacy and safety of intense focused ultrasound for skin rejuvenation: a systematic review and meta-analysis. Arch Dermatol Res. 2024;316(8):513.
15. Fatemi F, Najafian R, Sadri MA. New HIFU method: utilizing novel surgical and anatomical concepts yields outstanding results. J Cosmet Dermatol. 2024;21 https://doi.org/10.1111/jocd.16374.
16. Wood E, Gonzalez A, Almukhtar R, Fletcher L, Fabi S. Comparing the safety and effectiveness of microfocused ultrasound: standard vs targeted tissue protocol in lifting and tightening the lower face and upper neck. J Drugs Dermatol. 2024;23(4) https://doi.org/10.36849/JDD.7647.
17. Koza E, Ma M, Ahmed A, Haq M, Shi VJ, Whiting D, Paghdal K, Weil A, Voravutinon N, Roongpisuthipong W, Poon E, Alam M. A comparison of 1550-nm fractional Photothermolysis versus intense focused ultrasound for treatment of periorbital wrinkles: a randomized control trial. J Am Acad Dermatol. 2024;9:S0190-9622(24)00051-3. https://doi.org/10.1016/j.jaad.2023.12.047.

18. Al-Omair A, Bukhari A. Patient satisfaction of microfocused ultrasound treatments on the face and neck laxity: a narrative review. J Cosmet Dermatol. 2023;22(10):2671–6.
19. Casabona G, Kaye K, Cotofana S, Davidovic K, Alfertshofer M, Freytag L. Histological effects of a combined collagen stimulation procedure consisting of microfocused ultrasound, soft tissue filler, and Ca-HA injections. J Cosmet Dermatol. 2023;22(6):1724–30.
20. Chan K, Lee A, Lee C. Pilot study of microfocused ultrasound, Incobotulinum toxin, and calcium hydroxyapatite in triple therapy for skin tightening after weight loss. Cosmet. 2023;10:168. https://doi.org/10.3390/cosmetics10060168.
21. Alizadeh Z, Halabchi F, Bodaghabadi Z, Zarandi MM, Abolhasani M, Seifi V, Khazaei R, Ghanadi S, Mazaheri R, Tabesh MR. Non-invasive body contouring technologies: an updated narrative review. Aesth Plast Surg. 2024;48(4):659–79.
22. Park JY, Youn S, Hong W, Lee KC, Kim I. Treatment protocol on using microfocused ultrasound with visualization for skin quality improvement: the Korean experience. Plast Reconstr Surg Glob Open. 2023;11(5):e5029.
23. Park JY, Lee JS, Lee SR, Lee DH. Combined treatment with micro-focused ultrasound with visualization and intradermal Incobotulinumtoxin-a for enlarged facial pores: a retrospective study in Asians. Clin Cosmet Investig Dermatol. 2023;16:1249–55.
24. Ling J, Zhao H. A systematic review and meta-analysis of the clinical efficacy and patients' satisfaction of micro-focused ultrasound (MFU) treatment for facial rejuvenation and tightening. Aesth Plast Surg. 2023;47(5):1806–23.
25. Dodero D, Frascani F, Angelucci M, Bernabei G, Merlo E, et al. Histological modifications of postmenopausal vaginal mucosa after regenerative solid state laser treatment: a multicenter study. Int J Womens Health Wellness. 2019;5:099.
26. Longo L, Dell'Avanzato R, Longo D. Endolift and multi wavelenght laser photobiomodulation: a randomized controlled trial study on 96 subjects trating skin laxity for the lower third of the face. Laser Ther. 2024;29(2):115–20.
27. Hirokawa M, Kurihara N. Comparison of bare-tip and radial fiber in Endovenous laser ablation with 1470 nm diode laser. Ann Vasc Dis. 2014;7(3):239–45.
28. Doyle A, Looi I, Chu P. Microfocused ultrasound with visualization (MFU-V) and Hyperdilute calcium Hydroxylapatite (CaHA-CMC) of the lower face and Submentum to treat skin laxity: a pilot study demonstrating superiority of MFU-V first followed by Hyperdilute CaHA-CMC. Aesthet Surg J. 2024;7:sjae226.
29. Tsukahara K, Tamatsu Y, Sugawara Y, Shimada K. Relationship between the depth of facial wrinkles and the density of the retinacula cutis. Arch Dermatol. 2012;148(1):39–46.
30. Reissis Y, Garcìa-Gareta E, Korda M, Blunn GW, Hua J. The effect of temperature on the viability of human mesenchymal stem cells. Stem Cell Res Ther. 2013;4(6):139.
31. Liang H, Sheng W, Guan L, Qin F, Guo T. Effects of temperature on cell viability and chemosensitivity in human dermal fibroblasts. J Oncol Lett. 2017;
32. Nilforoushzadeh MA, Heidari-Kharaji M, Fakhim T, Hanifnia A, Nouri M, Roohaninasab M. Endolift laser for jowl fat reduction: clinical evaluation and biometric measurement. Lasers Med Sci. 2022;37(5):2397–401.
33. Nilforoushzadeh MA, Fakhim T, Heidari-Kharaji M, et al. Endolift laser an effective treatment modality for forehead wrinkles and frown line. J Cosmet Dermatol. 2022;21(6):2463–8.
34. Badin A, Moraes L, Gondek L, et al. Laser lipolysis: flaccidity under control. Aesth Plast Surg. 2002;26:335–9.
35. Scrimali L, Lomeo G, Dell'Avanzato R, Crippa A. Endolaser soft-lift: a new approach on body contouring. Perspectives and suggestions. Eur J Aesth Med Dermatol. 2013;3(3):86–90.
36. Roberto M. Endolift for the best face and body non-surgical lifting. Plast Surg Case Stud. 2020;1:1–2.
37. Dell'Avanzato R. Endolift® laser: survey over 10 experienced Italian doctors. In: 38th Italian Society of Aesthetic Medicine Congress (SIME); 2017.

38. Li K, Nicoli F, Xi WJ, Zhang Z, Cui C, Al- Mousawi A, Balzani A, Tong Y, Zhang Y. The 1470 nm diode laser with an intralesional fiber device: a proposed solution for the treatment of inflamed and infected keloids. Burns Trauma. 2019;15(7):5.
39. Brightman L, Brauer JA, et al. Ablative and fractional ablative lasers. Dermatol Clin. 2009;27(4):479–89.
40. Polder K, Friedman P, Feetham J, et al. Nonablative fractional diode laser resurfacing (1440 nm and 1927 nm) for Photoaged skin. Dermatol Surg. 2025;51(1):52–7.
41. Feras Marqa M, Mordon S. Laser fractional photothermolysis of the skin: numerical simulation of microthermal zones. J Cosmet Laser Ther. 2014;16(2):57–65.

Chapter 5
Autologous Regenerative Therapy in Baldness Surgery

Further, the role of extracellular vesicles, particularly exosomes, in modulating the hair growth cycle is discussed, alongside the potential of conditioned medium (CM) from stem cells as a paracrine therapeutic tool. The chapter also evaluates the controversial role of platelet-rich plasma (PRP) in hair regeneration, discussing the inconsistencies in clinical outcomes due to protocol variations.

The interplay between dermal white adipose tissue (dWAT) and hair follicle stem cells (HFSCs) is examined in-depth, shedding light on how adipose-derived growth factors influence hair follicle cycling, regeneration, and stability. The regenerative potential of stem cell-based therapies is framed within the broader context of tissue engineering and future possibilities in follicular bioengineering and neo-folliculogenesis.

Clinical case studies reinforce the efficacy of combining ART with hair transplantation, demonstrating improved graft survival, postoperative recovery, and long-term follicular density. By bridging the gap between traditional and regenerative approaches, this chapter presents a paradigm shift in hair restoration surgery, offering a more holistic and biologically sustainable solution for hair loss management.

5.1 A Brief Premise

Traditional medical therapies for androgenetic alopecia (AGA) aim to stabilize or improve the clinical condition.

Both objectives are important in addressing a progressive condition.

In modern care practices, slowing or halting the progression of a disease is the "achievable" and shared approach to assisting patients. In a broader context, neurologists today strive to halt the progression of Alzheimer's or Parkinson's with

D. Piccolo et al., *A.R.T. Autologous Regenerative Therapy in Aesthetic Medicine*, https://doi.org/10.1007/978-3-032-05835-5_5

therapies, while oncologists, in their field, use chemotherapy to slow or "chronicise" the progression of a tumor.

There is a stark difference in patient perception between these severe diseases, which pose a significant threat to the quality and duration of life, and a "social" condition like AGA. Patients tend to accept the limitations of treatments when faced with severe illnesses, whereas, when it comes to hair, their desires often exceed reasonableness.

Patients often fail to grasp the passage of time and the inevitability of aging, and some doctors also overlook the challenges of implementing approaches that promote long-lasting well-being.

Hair loss surgery has achieved significant advancements and successes. The recent US elections demonstrate that a surgical contribution to a successful image can make a difference. The four-time hair transplant recipient Donald Trump, and Elon Musk, who underwent two transplants, convinced the electorate that they were the right people to lead the country. Regardless of their numerous qualities, it is fair to question whether this would have occurred if their image had reflected their genetic predispositions.

Maintaining a youthful image, including a full head of hair, was certainly one of the concerns of the newly elected US president. His personal physician has openly acknowledged his use of finasteride to help maintain the outcomes of the surgical interventions performed.

"Maintaining results" is not just a keyword for public figures but resonates with the vast majority of individuals who view AGA as a distressing condition.

The dualism between traditional care and hair loss surgery, which has dominated the therapeutic landscape for alopecia in recent decades by offering two distinct yet often complementary approaches, may prove insufficient for various patient categories. Regenerative autologous therapy fits into this context. We aim to demonstrate that this option can no longer be considered superfluous, given some discrepancies that we will highlight between the ideal and actual conditions of each patient.

It should come as no surprise that this chapter is written by a surgeon and that this book on regenerative medicine includes many surgeons at the forefront. Many studies and much research are driven by situations where one feels constrained, and this historical moment presents many reasons to seek something new. A brief digression to clarify.

When a new building is delivered, the new homeowners benefit from a 10-year warranty, a mandatory insurance that the builder takes out to protect the buyer's investment. Similarly, car manufacturers offer 5 or more years of warranty on their vehicles. In both cases, manufacturers stand behind their product.

Similarly, a surgeon performing an operation, while not "guaranteeing" the outcome to the patient in a strict sense, still feels a responsibility for the long-term success. Accurate preoperative analyses and careful intervention planning are the main rules adopted to minimize risks and variability in outcomes. If this is the "mantra" for every surgeon, it becomes even more crucial in aesthetic surgery, where the results are immediately visible and judged by the patient.

Likewise, in trichology, there is generally no comparison between the burden a dermatologist experiences when their therapy for AGA fails and the weight a surgeon feels when a patient, years after a transplant, reports a loss of results. What drives this difference is not the doctor's sensitivity but the patient's expectations generated by their investment and future scenarios. Thus, a new and different therapeutic attempt is often well-received by a patient who, if dissatisfied, can switch doctors; conversely, a surgical failure leaves them with limited prospects, often only a new autotransplant that further "consumes" a limited donor area.

There are countless nuances and potential responsibilities, but one fact remains: in the "global market" of hair loss surgery, which is growing exponentially, there is a certain scenario where, in the years to come, many transplanted patients will experience result instability and seek alternative therapies.

Doctors must therefore move away from the tendency to simplify complex choices, from patients' desire to categorize everything into binary options, and learn to manage a reality that is often much more nuanced and complex.

People seek solutions that reflect not only their physical appearance but also their sense of identity and social success. Clinicians must work with patients' circumstances and limitations, aiming to optimize results in the best way possible. It is therefore important to adopt an integrated approach in medicine, taking into account the various dimensions of health and well-being as well as the complexity of human experiences.

5.2 Introduction

Regenerative medicine is currently one of the most relevant topics in the treatment of AGA, a condition that has always caused and continues to cause distress and concern. In recent decades, pharmacological therapies and hair loss surgery have achieved increasing success, but with significant limitations. Pharmacological treatments can vary in effectiveness, with side effects and challenges in maintaining long-term therapy. The duration of the results from hair transplantation, on the other hand, depends on subjective factors, such as the quality of the procedure, which can still vary significantly based on the experience and precision of the entire team, as well as objective factors related to the quality of the donor area.

Autologous Regenerative Therapy offers a new therapeutic perspective, focusing on tissue regeneration and the improvement of the perifollicular macroenvironment, i.e., the biological context surrounding the hair follicle that supports hair growth. This macroenvironment includes cells, tissues, blood vessels, nerves, and molecules that interact with each other and directly influence the health, growth, and life cycle of the hair follicle.

In recent years, hair transplantation techniques have evolved, focusing on reducing follicular trauma during extraction and implantation. However, attention has mainly been on the procedure itself, often overlooking the importance of the perifollicular environment, which, as numerous studies show, plays a crucial role in the progression of AGA and the success of the transplant. The consideration of this environment as a passive element and the lack of specific therapeutic approaches represent a significant limitation of traditional techniques.

Simple observations from folk traditions suggest that techniques such as scalp massage, microneedling, and practices like yoga promote hair regrowth by improving circulation, scalp health, and reducing stress. On the other end of the spectrum, more complex observations focusing on pretreatment of scarred areas before a transplant—which we will explore in the next chapter—demonstrate the importance of a healthy environment for follicular growth.

In this chapter, we will thoroughly examine the context in which regenerative methods are placed within hair loss surgery. We will explore the scientific foundations of the clinical use of adipose stem cells and present our case studies with the guided SEFFI method in relation to hair transplantation, thus offering a comprehensive and integrated view of the subject.

5.3 First Paragraph: Androgenetic Alopecia

5.3.1 Functional Anatomy of the Hair Follicle

The hair follicle is a unique structure of the skin that produces the cells destined to form the hair. It is an epidermal invagination shaped like a sac, with a deep dilation called the bulb, which narrows as it moves upwards towards the opening on the epidermis. Along with the hair and the associated sebaceous gland, the follicle forms the pilosebaceous unit. It consists of an epithelial part, the epidermis of the follicle, and a connective tissue part, enveloped by the glassy membrane, which connects to the surrounding dermis.

The epidermis of the follicle, which forms the walls of the follicle itself, is made up of several stacked epithelial layers. This stratification changes from the deeper part to the more superficial part, known as the collar. The boundary between these two areas is defined by the opening of the associated sebaceous gland and the insertion point of the arrector pili muscle (Fig. 5.1).

The hair bulb is the deepest part of the follicle, and in longer follicles, it can reach the hypodermis. It is the area responsible for generating all the components of the hair and hair follicle. The dermal papilla (DP) is a small structure located at the base of the hair follicle, within the bulb. It is made up of cells and connective tissue and is rich in blood vessels, which provide essential nutrients for hair growth. The papilla is crucial for regulating the hair cycle due to its ability to send signals and growth factors [1].

Surrounding the dermal papilla is the part of the bulb responsible for generating the hair, consisting of a thick layer of polyhedral and undifferentiated epithelial cells, forming the bulb matrix. These cells, similarly to the basal layer of the epidermis, reproduce and differentiate into keratinocytes that form the three structures of the hair: the medulla, cortex, and cuticle. Just above the matrix is the differentiation

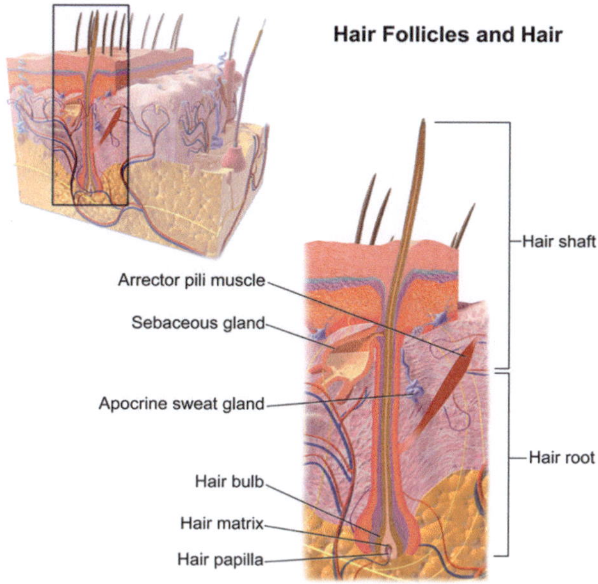

Fig. 5.1 Cross-sectional anatomy of human skin and hair follicle, illustrating key structures involved in hair growth and skin function, including the epidermis, dermis, sebaceous gland, arrector pili muscle, and components of the hair follicle such as the inner and outer root sheaths, matrix cells, and hair papilla

layer, the keratogenous zone, where the keratinization process of cells takes place, leading to the formation of the hair. Melanocytes, also present in this region, are arranged radially around the papilla and extend their projections through the first rows of overlying cells [1].

The normal hair growth cycle consists of four phases: anagen (growth), catagen (involution), telogen (rest), and exogen (shedding). In the anatomical area beneath the arrector pili muscle is the bulge, a reservoir of stem cells. At the end of each cycle, during the catagen phase, under the influence of specific biochemical signals, these stem cells migrate to the dermal papilla, giving rise to a new hair [1, 2]. During this phase, the cells of the sac produce estrone, cortisone, and glycogen, activating the stem cell compartment. These three elements are essential for ensuring the quality of the following anagen phase. Even a small metabolic error during the catagen phase can lead to hair miniaturization in the subsequent anagen phase. The telogen phase is activated by the hormone dihydrotestosterone (DHT).

5.3.1.1 Cells and Structures Involved in Follicular Regeneration

Here, we briefly mention the cells and structures involved in the regeneration of the hair follicle under physiological conditions. In recent years, we have come to understand that this cyclical phenomenon of hair growth does not solely depend on the stem cells of the hair follicle (HFSCs) but also involves coordinated interactions with surrounding cells and structures [9].

We know that distinct populations of HFSCs exist in the bulge and dermal papilla, and during the hair cycle, HFSCs interact with specialized dermal fibroblasts from the dermal sheath papilla, allowing the well-known transitions between each phase of the follicular cycle (Fig. 5.2).

This perspective highlights the crucial role of the structures that make up the follicle, but looking at a broader context that more closely aligns with the focus of this chapter, the entire scalp with its follicles is like a miniature rainforest, where every element plays a crucial role in supporting the life and vitality of the entire ecosystem.

Using a metaphor, we could say that just as the Amazon rainforest could not exist without the valuable contribution of the phosphorus-rich Sahara sand carried by the winds from the Atlantic Ocean, the hair follicle requires the support of seemingly small, distant, and only apparently insignificant components to survive. The blood and lymphatic vessels surrounding the follicle are like the waterways nourishing the

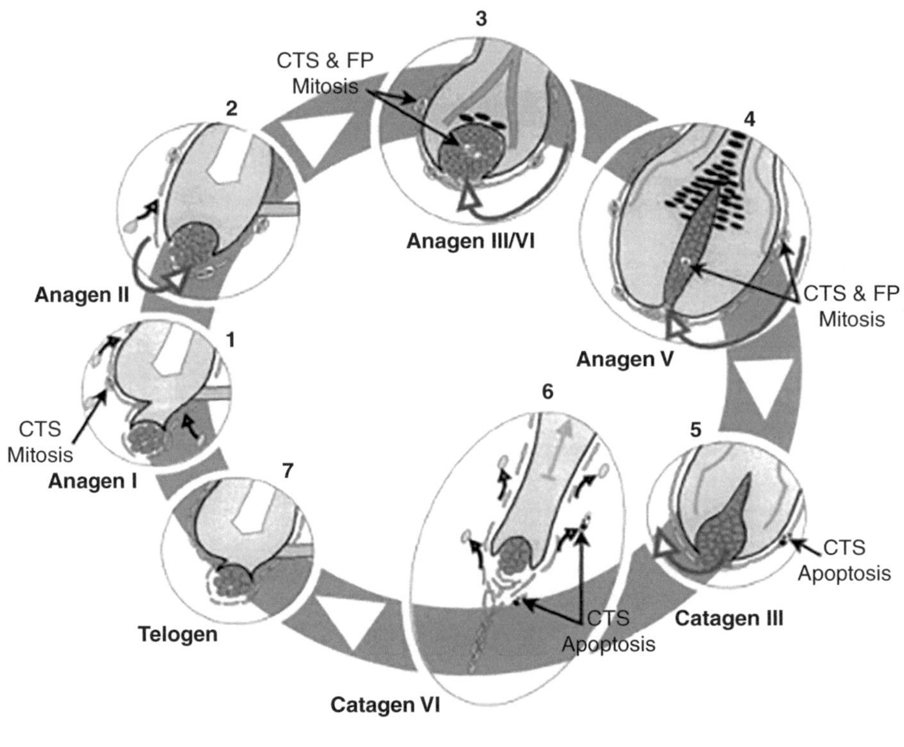

Fig. 5.2 Diagram showing the hair growth cycle with seven stages illustrated in a circular layout [14]. The stages include
Anagen I—initiation of growth with CTS (connective tissue sheath) mitosis
Anagen II—continued growth phase
Anagen III/VI—active mitosis in CTS and FP (follicular papilla)
Anagen V—further follicle development with CTS and FP mitosis
Catagen III—beginning of regression with CTS apoptosis
Catagen VI—continued regression with follicle shrinking and CTS apoptosis
Telogen—resting phase with no active growth
The cycle represents the dynamic process of hair growth, regression, rest, and renewal, with cellular events like mitosis and apoptosis indicated at various stages

forest, modulating their flow according to the different stages of the hair growth cycle. The rich nerve plexus enveloping the follicle is akin to a dense network of paths that allow communication and coordination between the different parts of the ecosystem. Mediators such as substance P and noradrenaline act as messages sent between the inhabitants of this miniature forest, influencing hair growth.

However, the real secret of this follicular ecosystem lies in the communication that takes place at the cellular level, and, just as microorganisms and insects in the Amazon rainforest exchange information through subtle "threads" of communication, in the scalp, extracellular vesicles, called exosomes [12], perform this crucial role, carrying targeted messages between follicle cells like small couriers delivering vital instructions for maintaining and growing hair.

It is the small but essential contributions that make the difference in this delicate ecosystem. Understanding the interdependence of the cells and structures involved in the hair growth cycle, as well as the mediators of intercellular signaling, is key to understanding hair regeneration. This is an undertaking that has just begun and presents many avenues for clinicians and researchers, not only for the benefit of patients with androgenetic alopecia but also for those suffering from other alopecic conditions.

5.3.2 Molecular Mechanisms Underlying Androgenetic Alopecia

As explained in the previous chapter, AGA is a widespread chronic condition that is genetically determined, characterized by the progressive involution of hair follicles and the hair that arises from them. This involution is caused by androgens, particularly by dihydrotestosterone (DHT), the most potent androgen hormone, which is considered a pure androgen because it does not convert into estrogens [4, 5].

DHT does not only act endocrinologically but also, and above all, para-cranially. In fact, DHT is mainly formed in peripheral tissues, where it exerts its effects. Specifically, testosterone is converted into DHT by the enzyme 5-alpha reductase [4] in target tissues, and this localized (paracrine) synthesis results in chemical messengers being produced and released by diffusion, which modify the physiology of surrounding cells.

In AGA, androgen-sensitive hair follicles have highly sensitive but low-capacity receptors in the dermal papilla cells. When androgens bind to these receptors, they directly influence target cells, such as keratinocytes. There are no significant hormonal differences between a person affected by AGA and a healthy individual; however, in predisposed subjects, even a low concentration of androgens is enough to bind to a substantial number of receptors, triggering the condition.

In the pathogenesis of AGA, higher levels of the enzyme 5-alpha reductase type II are found, which is responsible for the intracellular conversion of testosterone into DHT. This triggers a cascade of events that deregulate the follicular cycle, affecting gene transcription, energy synthesis, and cellular energy production. This is how an individual's genetic message becomes their phenotype.

It may be difficult to understand the paradoxical effect of this hormone and why androgens stimulate hair growth on most of the body but inhibit it on the scalp. The difference in response lies in the signal transduction downstream of receptor binding. In other words, although DHT binds to the receptors with the same affinity, scalp cells and those from other parts of the body, such as the beard, respond differently, resulting in opposite outcomes.

Years of research have provided complex insights, but in this context, it suffices to remember that AGA has a multigenic and multifactorial etiology. This indicates that its development is not only regulated by more than one gene [4, 5], but also by other factors that may play a co-causal role, such as changes in weight, poor diet, hypothyroidism, stress states, etc.

The extreme variability of penetrance results in a range of phenotypes, from mild forms consisting of a slight recession of the front hairline to severe and obvious forms such as complete alopecia of the central scalp area.

What is most likely inherited is the follicular enzymatic makeup involved in the conversion and uptake of androgens, more precisely: the two isoenzymatic forms of 5α reductase (type 1 and type 2), P450 aromatase, 3α HSD, 17α HSD, and the cytosolic androgen receptor [4].

The different clinical presentations of AGA in men and women (patterns) can be explained by the differing levels of androgen receptors and androgen-converting enzymes present in various areas of the scalp in men and women. These changes reflect the progressive transformation of terminal hairs into vellus hairs, which is typical of androgenetic alopecia (Fig. 5.3).

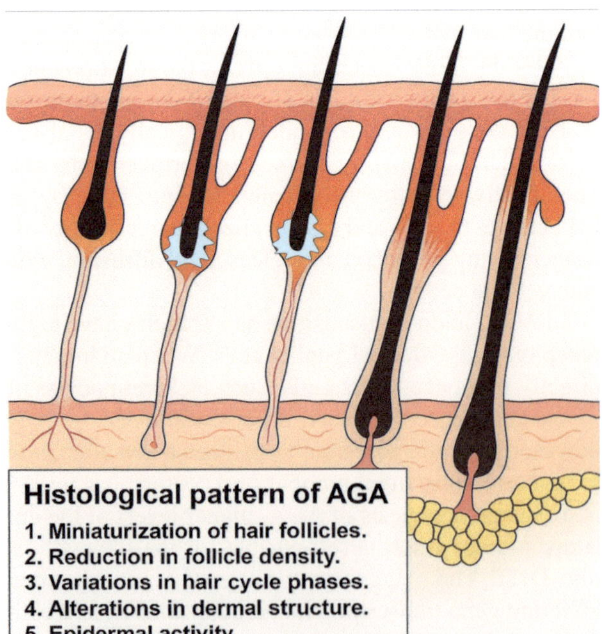

Fig. 5.3 Histological pattern of androgenetic alopecia (AGA), illustrating progressive follicular miniaturization, reduced follicle density, altered hair cycle phases, dermal structural changes, and modifications in epidermal activity. These features are characteristic of AGA-related hair loss

Although hair density tends to decrease with age following a given pattern, it is not possible to predict the extent of hair loss over time in a young person with early signs of androgenetic alopecia. Generally, those who begin to lose hair in their second decade of life are those in whom alopecia will progress the most.

In some men, hair loss may occur even later, at the end of the third or during the fourth decade of life. In women affected by AGA, it is much more likely that hair loss will first appear between the ages of 30 and 40 (compared to 20 to 30 in men). The condition may appear in association with hormonal changes, such as the initiation or cessation of contraceptive pill use, the postpartum period, post-pubertal period, or during the perimenopausal and postmenopausal phases, due to significant weight changes that alter hormonal balance, or conditions that cause hormonal imbalances such as polycystic ovary syndrome (PCOS).

Bitemporal recession occurs during sexual maturation in 80% of women (as in most men), but it is generally much less prominent than in men.

5.3.3 Diagnosis of AGA: Dermatoscopic, Clinical, and Trichoscopic Evaluation

The diagnosis of AGA in clinical practice is based on three main elements: patient history, clinical assessment, and videodermatoscopic analysis.

The latter, in particular, allows for a detailed evaluation of the nature of hair loss, the degree of follicular miniaturization, and the possible presence of perifollicular erythema and/or scales. This examination is crucial for distinguishing AGA from other conditions that can cause hair loss.

The earliest and most diagnostic sign of AGA is a change in the diameter of the hair shafts, with a reduction of more than 20% compared to surrounding hairs. Yellow dots may also be present, indicative of hypertrophied sebaceous glands, often associated with perifollicular inflammation.

5.4 The Pharmacological Treatment of AGA

The goal of pharmacological therapy for the treatment of androgenetic alopecia (AGA) is to reduce the progression of hair loss, stabilize or reverse the clinical condition, and, when necessary, create optimal conditions for further improvement through a hair transplant procedure. The available pharmacological therapies not only stop hair loss but can also increase the thickness of existing hair and reduce thinning. Currently, only two pharmacological treatments are FDA-approved for the treatment of AGA: minoxidil and finasteride.

5.4.1 Main Pharmacological Treatments and Mechanisms of Action of Finasteride and Minoxidil

Minoxidil is a topical vasodilator that acts as an activator of ATP-dependent potassium channels. It is an effective therapeutic agent in the treatment of androgenetic alopecia in both sexes. The mechanism of action of minoxidil is primarily based on modulating the hair follicle growth cycle, promoting the transition from the telogen phase (resting phase) to the anagen phase (active growth phase).

The multifactorial mechanism of action of minoxidil includes the following:

- Induction of the anagen phase: it stimulates the conversion of follicles from the telogen phase to the anagen phase, thereby initiating a new growth cycle.
- Prolongation of the anagen phase: it extends the duration of the active growth phase, promoting longer hair.
- Improvement of microcirculation: it induces vasodilation in the microcirculation of the scalp, optimizing blood supply and the availability of nutrients and growth factors to the hair follicles.

Stimulation of dermal papilla cells: it promotes the proliferation and differentiation of cells in the dermal papilla, an essential component that regulates the hair growth cycle.

Clinical studies have demonstrated the effectiveness of minoxidil in the treatment of AGA in both men and women. At the molecular level, minoxidil activates potassium channels, triggering a cascade of cellular events, including increased expression of vascular endothelial growth factor (VEGF) and the stimulation of prostaglandin synthesis, both crucial for promoting hair growth. In conclusion, minoxidil is a scientifically supported therapeutic treatment for AGA, thanks to its multimodal mechanism of action, which positively influences various aspects of hair follicle physiology [6].

Finasteride dihydrotestosterone (DHT), a potent metabolite of testosterone, is the primary hormone involved in the pathogenesis of male androgenetic alopecia. DHT selectively acts on androgen-responsive hair follicles, primarily in the frontotemporal and vertex areas, where it causes hair miniaturization. Prolonged exposure to DHT interacts with androgen receptors on the hair follicles, triggering a progressive reduction in the diameter of the hair shaft and shortening the anagen phase. These effects primarily manifest in the dermal papilla, where the concentration of androgen receptors is highest.

Although the precise mechanism of miniaturization is not yet fully understood, it has been shown that DHT, once bound to its receptors, activates genes responsible for androgen-responsive proteins, such as TGFB1 (which codes for TGF-β1), TGFB2 (for TGF-β2), DKK-1, and IL-6. These genes inhibit keratinocyte proliferation (TGF-β1), induce keratinocyte apoptosis (TGF-β2 and DKK-1), or promote the onset of the catagen phase (IL-6) [1, 3].

Finasteride works by inhibiting the enzyme 5-alpha reductase type II, which is responsible for converting testosterone into DHT. The competitive binding of finasteride to the enzyme's active site blocks its catalytic activity [7], reducing DHT levels in target tissues, such as the scalp. By blocking DHT synthesis, finasteride halts hair loss and, in some cases, stimulates the regrowth of hair in miniaturized follicles [7]. Its efficacy is well documented in men, but it can also be used in women with AGA under careful medical supervision.

5.4.2 Role of Pharmacological Therapy in the Context of Hair Transplantation

Pharmacological treatment of androgenetic alopecia not only counters the progression of the disease but also improves the quality and density of existing hair. The application of minoxidil (topical) and finasteride (systemic) represents an effective strategy for improving or consolidating clinical outcomes, both in the preoperative and postoperative periods of a hair transplant procedure.

Both drugs must be used consistently to maintain the benefits achieved, as their discontinuation leads to a progressive and rapid loss of results. Although side effects are often overstated, these medications are associated with common concerns among patients, exacerbated by pharmacophobia.

Before a transplant, pharmacological therapy helps stabilize hair loss, increase the caliber of hair in the donor area, and reduce the size of the recipient area, creating a solid foundation for the success of the procedure. Preoperative therapeutic results help reinforce the patient's belief that the therapy is a crucial element for the effectiveness of the intervention, improving their perception of the importance of self-care in the long term. An additional benefit of therapy is the possibility of reducing the number of follicular units needed for the transplant, limiting the extent of graft harvesting from the donor area.

After the transplant, pharmacological therapy helps reduce the loss of native hair and maintain long-term results, enhancing the overall effectiveness of the treatment. Therefore, an integrated approach combining pharmacological treatments and surgical interventions represents the optimal solution for managing AGA, ensuring the best long-term results.

5.5 Surgical Treatment of AGA

The follicular unit extraction (FUE) technique is a minimally invasive hair transplant procedure that allows for the extraction of individual follicular units (FUs) directly from the patient's scalp, from the so-called donor area, to be implanted into thinning or balding areas, the recipient areas.

This brief paragraph does not aim to describe the technique in detail but to highlight some particularities of the operation that clarify the current context.

5.5.1 *Improvements in Surgical Techniques for Hair Transplantation*

FUE is a manual extraction technique, using a hollow cylindrical punch (manual or motorized) to collect the follicular units. These units are then extracted using forceps. While the technique employs relatively simple instruments, several key innovations have greatly enhanced the surgeon's effectiveness, enabling reproducible results that are less influenced by the individual characteristics of the patients.

Taking a critical look at the entire process, the main innovations have focused on the instrumentation and the preparation of the operators. It is crucial to avoid damage to the follicular structures, and this remains the best guarantee of success and longevity in hair transplantation. Other phases, such as the preservation of the follicles during their extracorporeal period, have also received attention, and various preservation solutions seem to ensure adequate protection for the follicles before reimplantation. All these advancements have improved the quality and reliability of results, contributing to the growing demand for hair transplants.

It can, therefore, be stated that a hair transplant procedure can yield satisfactory and natural results, with minimal scarring in the donor area.

However, it is important to highlight that hair transplantation is not a universal solution for all types of alopecia. In particular, some women and, less frequently, some men present with forms of diffuse alopecia (DUPA—Diffuse Unpatterned Alopecia) characterized by widespread miniaturization of hair follicles across the entire scalp. In these cases, even the donor area is involved in the miniaturization process, reducing the availability of suitable follicular units for transplantation. As a result, hair transplantation is not a suitable therapeutic option for these patients. Such conditions require alternative, personalized approaches.

5.5.2 *Instability and Loss of Surgical Results*

Many surgeons are reluctant to admit that hair transplantation is a procedure with variables that are not always controllable, much like most surgical interventions. A critical factor to consider is the potential long-term instability of the results obtained in some categories of patients. This instability may stem from various factors, related both to the vitality of the transplanted hair and the health of the native hair in the recipient area.

The loss of native hair in the treated area may be due to the following:

Lack of maintenance therapy: The absence of pharmacological treatment to counteract the progression of AGA can lead to the progressive loss of non-transplanted hair.

Trauma to native follicles: Mechanical trauma to the preexisting follicles during the insertion of grafts can lead to their loss.

Aging: The physiological follicular miniaturization related to age, typically after 50 years [13], can affect both transplanted and native hair.

Regarding the loss of transplanted hair, this may be due to the following:

Failure to take root: A percentage of grafts may not take properly and, consequently, will not grow. This failure to take root can be related to intrinsic factors of the patient or technical complications during the procedure.

Alteration in hair quality: After regrowth in the recipient area, transplanted hair may have inferior qualitative characteristics compared to the donor site, and their unexpected thinning, sometimes progressive over time, can lead to a result below expectations.

Although transplanted hair is resistant to the molecular mechanisms of AGA, thinning or miniaturization over time cannot be completely excluded due to the following:

Harvesting from unsuitable areas: Using follicular units harvested from donor areas with a high likelihood of follicular miniaturization (e.g., areas already affected by AGA) can compromise the longevity of the result.

Damage during harvesting: A high percentage of transection (the percentage of follicles damaged during extraction) compromises the vitality of the grafts and their potential for long-term growth. This damage may affect the stem cell reservoir in the follicular bulge.

Damage during manipulation: Damage to follicular units during the extraction, preservation, and implantation phases can reduce their vitality.

Aging: Aging may cause miniaturization even of transplanted hair.

The previous two paragraphs provide a quick overview of the many advantages but also the limitations of hair transplantation, procedures that rely on the varying resources of the donor area. It is also important to consider, among the potential challenges of this procedure, two psychosocial aspects: the possible ethical implications related to the professional aspirations of surgeons, and the complex motivations of patients, which may include unrealistic expectations or reluctance to change their lifestyle.

The surgeon's desire to achieve optimal results can sometimes lead to overly optimistic predictions, such as expectations of complete graft survival or perfect adherence to postoperative care by patients. Confidence in their technical abilities, especially in the early stages of their career, may lead some professionals to use the donor areas less conservatively. When results do not meet initial expectations, corrective options may prove limited and potentially unsatisfactory.

At the same time, patients may be influenced by unrealistic expectations, underestimating the limitations imposed by their specific pattern of baldness or overestimating their ability to adhere strictly to the postoperative therapeutic regimen. Often, the daily circumstances that interfere with the implementation of necessary care are not adequately considered, even in the presence of strong initial determination.

It is important to recognize that, in this field, as in many other areas of medicine, situations rarely present themselves in terms of absolute "black or white." Clinical reality is often characterized by nuances and complexities that require a nuanced and multidisciplinary approach. This awareness has led the field of hair restoration surgery to look with increasing interest towards regenerative medicine, seeking innovative solutions that can integrate and enhance traditional approaches.

5.6 Fourth Paragraph: New Therapeutic Possibilities for AGA

5.6.1 Introduction

How can regenerative medicine be placed in this clinical context? What regenerative medicine is currently available and for what purposes? Can it truly solve some of the issues described, and how?

To answer these questions, it is useful to examine the evolution of expectations and research in the field of regenerative medicine applied to AGA over the past few decades. The promise of a revolution in the treatment of baldness through hair follicle cloning, with optimistic predictions of concrete results within two decades, has characterized discussions in the sector since the late 1990s and early 2000s.

These expectations, fueled by early promising research on stem cells and their regenerative capacity, have largely proven unfounded. While basic research has led to a deeper understanding of the molecular mechanisms underlying hair growth and the potential of dermal papilla cells, the translation of these findings into effective and widely available clinical treatments has proven much more complex than expected.

5.6.2 Studies on the Dermal Papilla

The development of regenerative therapies focused on the ability of dermal papilla (DP) cells to induce biological processes capable of creating new follicles began many years ago. This group of specialized mesenchymal cells possesses stem cell characteristics and can be cultivated and expanded in vitro [12]. Research had promised effective outpatient treatments, and the companies involved, as well as investors and doctors, had hoped that the time for a major breakthrough was approaching.

The research conducted explored two main strands: the regeneration and rejuvenation of hair follicles.

"De novo" regeneration (a term indicating the formation of an entirely new structure or tissue) aims to create new hair follicles by inducing follicular ectogenesis in the overlying epidermis following the injection of DP cells.

In contrast, follicular rejuvenation techniques exploit the ability of DP cells injected into the scalp to migrate towards miniaturized follicles, restoring or enhancing the functionality of the dermal papilla and promoting the transition from vellus follicles (thin and depigmented) to terminal follicles.

Today, there is an improved understanding of the complex molecular mechanisms involved in hair follicle morphogenesis, but the translation of these results into an effective and reproducible clinical treatment remains a significant challenge. Injecting DP cells, even those derived from three-dimensional cultures, does not currently guarantee the "de novo" formation of a clinically relevant number of functional hair follicles. There are still major technological and biological obstacles to overcome, and wisely, companies have ceased to promise a definitive solution to baldness as being "just around the corner."

The next paragraph will outline an alternative therapeutic approach based on the use of the stromal vascular fraction derived from autologous adipose tissue, but before delving into the specifics of this technique, it is worth exploring the scientific foundations that link the dermal adipose compartment to the hair follicle.

5.6.3 Interaction Between dWAT and HFs

Dermal white adipose tissue (dWAT) and hair follicles (HFs) are interconnected structures that actively interact through paracrine signaling, synchronizing their activities during the hair cycle. These communications influence the cyclical behavior and volume of both tissues in both physiological and pathological conditions, with important information derived from the latter. For example, following skin wound, regenerating hair follicles secrete factors that reprogram fibroblasts into adipocytes, contributing to adipose tissue regeneration. In the case of bacterial infection, both dermal adipocytes and hair follicles rapidly activate an innate immune response.

dWAT, working closely with HFs, performs crucial immunoregulatory and regenerative functions for the skin. These aspects highlight the importance of the interaction between dWAT and HFs in ensuring adequate thermoregulation, immune protection, and skin regeneration.

This functional coupling of the hair follicle (HF) and dermal white adipose tissue (dWAT) occurs towards the end of embryonic development after an initial distinct development. We can consider these tissues as separate, yet physiologically interconnected structures.

Common fibroblast progenitors populate the future dermal compartment. In the upper dermis, these progenitors differentiate into cells of the developing hair follicle, while those located in the lower part give rise to reticular progenitors, some of which then transform into adipogenic precursors of dWAT.

Compared to subcutaneous (sWAT) and visceral (vWAT) adipose tissue, which form amorphous and less organized masses, dWAT has a fine and well-structured microanatomy.

A distinctive aspect of dWAT is its interaction with hair follicles (HFs), which varies depending on their cyclical activity. During the anagen phase, hair follicles penetrate into dWAT, which thickens and wraps around the base of the mature hair follicle, protecting it from trauma like the cushioning of fat pads in joints, thus promoting growth. Conversely, when the hair follicles shorten during the telogen phase, dWAT thins. Remodeling of dWAT is the result of periodic macro-environmental signals modulating adipogenesis, originating from various follicle cell types, in addition to systemic metabolic stimuli.

Embryological development, anatomical position, and the organization of dWAT make it a unique adipose compartment that, while structurally adjacent to sWAT in humans, has a distinctive cellular composition and metabolic activity compared to it.

The interaction between dermal white adipose tissue (dWAT) and hair follicles (HFs) manifests through a series of interconnected biological processes that mutually support the functionality and health of both structures:

- Extracellular Matrix (ECM) Remodeling: dWAT participates in remodeling the skin's extracellular matrix, facilitating the expansion of HFs during anagen.
- Source of Nutrients: It provides free fatty acids (FFA) as nourishment for the energetically demanding growth of HFs.
- Capillary Network Remodeling: Both dWAT and HFs are rich in capillaries and lymphatic vessels, and it is plausible that these vessels may serve as conduits for the exchange of essential nutrients and hormones between these two structures.
- Thermoregulation: dWAT acts as a thermal insulator, working with HFs to trap warm air, thus contributing to thermoregulation.

These synergistic functions between dWAT and HFs indicate a complex system of interactions that support the health and optimal functioning of the skin, contributing to various aspects of its physiology and regeneration.

5.6.4 Molecular Mechanisms Between dWAT and HFs

The complexity of the interactions between dermal white adipose tissue (dWAT) and hair follicles (HFs) suggests intricate coordination capable of modulating not only the growth of adipose cells but also the hair cycle. In cellular biology, signaling mechanisms and pathways represent two fundamental concepts for understanding how cells communicate with each other and respond to external stimuli. Signaling mechanisms encompass the overall modes of cellular communication, involving the reception and transduction of signals through receptors and transduction molecules. Signaling pathways, on the other hand, are the specific routes through which signals are transmitted, culminating in biological effects.

5.6.4.1 Signaling Mechanisms

The interaction between dermal white adipose tissue (dWAT) and hair follicles (HFs) occurs through soluble signaling factors. During the anagen phase, HFs secrete ligands and antagonists for various signaling pathways involved in adipogenesis, such as BMP, WNT, PDGF, FGF, and Hedgehog. These secretions stimulate adipogenesis in dWAT, activating metabolic pathways that promote the differentiation of adipogenic precursors into mature adipocytes [3, 10, 11].

dWAT, in turn, releases ligands such as BMP (from the TGF-β superfamily), adipokines (such as leptin and adiponectin), and inflammatory cytokines, influencing the stem cells of HFs. BMP signaling is particularly important: HFs express BMP ligands that promote the development of dWAT (adipogenesis) through the activation of the nuclear receptor PPARγ [3].

5.6.4.2 Signaling Pathways

The interaction between dWAT and HFs is influenced by various signaling pathways:

- WNT Signaling: Plays an anti-adipogenic role and influences dWAT during the healing process.
- PDGF Signaling: Crucial for the induction of anagen and the maintenance of stem cells in HFs.
- Hedgehog Signaling: Promotes the proliferation of adipocyte precursors in dWAT during the anagen phase [3–10]

5.6.5 Role of Adipocytes in the Epithelial-Mesenchymal Interaction

The morphogenesis and cyclic regeneration of HFs depend on the epithelial-mesenchymal interaction (EMI), which sees intense and close cooperation between two cell populations: the epithelial cells of the bulge, which contain epidermal stem cells (Epi-SCs), and the mesenchymal cells of the dermal papilla (DP) [10].

Dermal white adipose tissue (dWAT), located just below the epidermis, plays a supportive role in this process. During EMI, the DPs release growth factors and molecular signals that influence the behavior of Epi-SCs, stimulating activities such as proliferation, migration, and differentiation. At the same time, Epi-SCs send back signals that regulate the functions and proliferation of DPs, which have a greater ability to differentiate into various cell types, creating a feedback loop that regulates the development and regeneration of tissue [10].

dWAT interacts with both to help regulate the hair cycle. Additionally, dWAT itself is a source of mesenchymal stem cells and secretes paracrine factors that can positively influence EMI, supporting hair regeneration and growth processes.

The dWAT not only provides structural support and nutrients but also secretes factors that promote communication between the epithelial and mesenchymal cells of the hair follicle. This support helps maintain a macroenvironment conducive to follicular regeneration and the progression of the hair cycle, thereby stimulating the growth phase (anagen).

In summary, the EMI is a coordinated interaction between the epithelial and mesenchymal cells, essential for the health and regeneration of hair follicles. Dermal white adipose tissue, with its ability to produce signals and provide cellular support, is an integral element of this process, contributing to overall function and hair regeneration.

5.7 Fifth Paragraph: Therapies Based on the Use of Stem Cells

5.7.1 Scientific Basis for the Clinical Use of Adipose Stem Cells

To understand the therapeutic strategies based on the use of stem cells for the regeneration of hair follicles (HFs), it is important to recall that in AGA, miniaturized follicles are characterized by a decrease in the number of cells within the dermal papilla (DP). Although the number of stem cells in the bulge (HFSC) is equivalent to that observed in terminal follicles, the number of progenitor cells is significantly reduced.

It is hypothesized that the driving force in AGA may have two origins:

- **Mesenchymal**, in the case of an intrinsic defect in the HFSCs (located in the bulge region), resulting in their inability to differentiate into progenitors
- **Epithelial**, in the case of altered or weakened signaling from the DP to the HFSCs in the bulge, initiating miniaturization (Fig. 5.4).

Currently, two therapeutic strategies have been proposed for the regeneration of HFs using stem cells, targeting both of the mechanisms mentioned above:

- Reverse the pathogenesis of hair loss by activating the bulge (HFSCs).
- Reverse the pathogenesis of hair loss by activating the DP.

The effectiveness of these therapeutic strategies is based on the mechanisms by which stem cells act in follicular regeneration. It is believed that stem cells can reverse the pathological mechanism primarily through paracrine actions. These cells secrete growth factors, cytokines, and other signals that activate hair follicle (HF) cells within their native niche, thereby contributing to their regeneration.

Another mode of action of stem cell therapies is direct differentiation. In this process, stem cells transform into specialized cells required for the reconstruction of hair follicles.

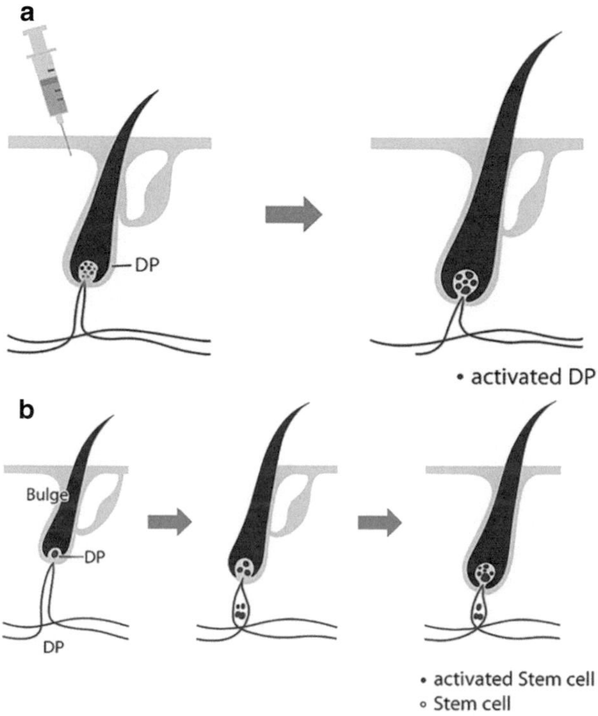

Fig. 5.4 Therapeutic strategies for hair follicle regeneration in androgenetic alopecia (AGA) using stem cells. (A) Targeting hair follicle stem cells (HFSCs) in the bulge area to reverse AGA. (B) Stimulating dermal papilla (DP) cells to promote hair regrowth. Both approaches aim to reactivate key regenerative components within the hair follicle [15]

These modes of action provide the scientific basis for the clinical use of adipose stem cells in follicular regeneration, offering potential innovative solutions to counteract the progression of AGA.

Within stem cell-based therapeutic strategies for follicular regeneration, it is important to examine the different application methods and their practical and clinical implications. With the advancement of regenerative medicine, therapies based on stem cells derived from various tissues, such as adipose tissue, bone marrow, follicle, and umbilical cord blood, have opened new pathways to address the challenges posed by conventional hair loss treatments.

These therapies, proposed for their potential to regenerate hair follicles, have been developed in laboratories with the idea of using a product, often heterologous, obtained from cells collected from donors and cultivated in vitro. Their high multipotent differentiation capacity and low immunogenic potential have sparked significant interest.

However, there are many legislative limitations surrounding stem cell therapies, and significant issues related to the costs of short storage durations, production, transportation, and specialized storage.

These technical and regulatory challenges have stimulated the search for alternative therapeutic approaches that can offer greater safety, efficacy, and economic sustainability, driving the scientific community toward exploring innovative solutions in the field of follicular regeneration. Consequently, interest in conditioned media (CM) has grown, given its lower susceptibility to allergic reactions and broader tolerability.

5.7.2 Conditioned Media

Therapies based on conditioned media derived from stem cells (CM) are widely explored in the clinical context as a promising therapy for hair follicle regeneration. CM is a culture medium in which stem cells have been grown, during which the cells release various bioactive molecules such as growth factors, cytokines, proteins, and other molecular signals that reflect their ability to communicate and influence other cells. The CM is then separated from the cells and referred to as the secretome, allowing for the bioactivity of the molecules and vesicles contained within it to be used in different contexts without the need to implant the cells themselves.

This approach helps avoid the potential risks associated with cell-based therapies, including rejection, tumor development, infections, surgical complications, and variability in results.

In AGA, where the ability of hair follicle stem cells to differentiate into progenitors is compromised, the aim of stem cell-derived CM is to restore the signals responsible for activating the hair follicle stem cells.

Although the clinical use of secretome derived from stem cells of the DP in hair loss treatment is in its early stages, many clinical studies have shown positive effects on hair regeneration.

Furthermore, CM from adipose stem cells, containing growth factors and cytokines, has been shown to increase hair thickness and density, stimulate the transition to anagen, and promote the proliferation of dermal papilla cells through specific signaling pathways. This is particularly relevant as, due to certain limitations of therapies based on the direct use of stem cells, the use of stem cell secretome is emerging as a promising new alternative for the treatment of hair loss. Studies, such as that by Gnecchi and colleagues, have revealed that mesenchymal stem cells (MSCs) exert their therapeutic effects primarily through the release of soluble factors (secretome), rather than through the integration of the cells themselves into damaged sites. This secretome-based approach avoids the potential risks associated with therapies focused on the direct use of cells, especially if heterologous.

Considering the inherently complex nature of hair regeneration, which involves the activation and differentiation of hair follicle stem cells, it is likely that a group of paracrine factors, rather than a single one, are responsible for the observed increase in hair growth and the number of follicles in the anagen phase. A group of paracrine factors, rather than a single factor, is likely responsible for promoting hair growth by increasing the number of follicles in anagen.

Larger samples and improved administration and evaluation methodologies, with longer follow-up periods, are needed.

5.7.3 Stromal Vascular Fraction (SVF) Derived from Adipose Tissue

The stromal vascular fraction (SVF) is a heterogeneous collection of noncultivated cells isolated from adipose tissue through minimal manipulation, centrifugation, filtration, and purification. These cells have attracted significant attention in various fields of regenerative medicine for their therapeutic potential, including in trichology.

It is hypothesized that the regenerative effect of SVF occurs through four main mechanisms:

- *Direct replacement*: Stem cells and progenitor cells in the SVF directly replace the host's progenitor cells.
- *Tissue rejuvenation*: SVF promotes tissue renewal by reversing the process of cellular differentiation.
- *Paracrine and endocrine effects*: Transplanted SVF contains trophic, angiogenic, and immunomodulatory factors that signal to nearby (paracrine) or distant (endocrine) cells.
- *Cellular combination*: SVF cells may combine with host cells.

In the context of hair growth, the importance of the relationship between hair follicles (HFs) and dermal white adipose tissue (DWAT) has been demonstrated. The precursor adipocytes present in the SVF activate hair follicle stem cells (HFSCs) by releasing growth factors such as PDGF, which can induce the transition of follicles from the telogen phase [3, 10].

Advantages and Considerations:

- The use of autologous stem cells reduces the risk of rejection.
- The risk of tumor development is low, except in cases of preexisting undiagnosed tumors.
- Standardized procedures reduce the risks of infections and complications.

It is important to note that this therapy is well-suited to complement traditional treatments, often yielding the best results when combined with them. Therefore, rather than being a "stand-alone" solution, SVF may restore the tissue's ability to respond to previously ineffective treatments on compromised tissue.

The stem cell fraction (ADSC) is part of the SVF derived from adipose tissue, and it is believed that the promoting effect on hair growth of ADSCs occurs through the enlargement and protection of the dermal papilla (DP). The paracrine factors secreted can:

Promote cell proliferation by activating signaling pathways such as Erk and Akt [10].

Remedy damage by neutralizing oxidative stress and protecting DP cells from the negative effects of androgens and free radicals.

Stimulate tissue regeneration by supporting the repair and maintenance of a healthy and functioning DP.

In conclusion, the cells from the stromal vascular fraction derived from adipose tissue represent a promising frontier in the therapy for hair loss, particularly for the treatment of androgenetic alopecia. Their regenerative potential, coupled with their ability to secrete growth factors and stimulate hair follicle stem cells, offers new therapeutic possibilities.

However, it is essential to emphasize that:

Research in this area is still evolving, and further controlled and randomized clinical studies are needed to fully confirm the efficacy and safety of these treatments.

The most effective approach seems to be integrating these innovative therapies with traditional treatments, rather than considering them as isolated solutions.

Treatment individualization, taking into account the patient's individual characteristics, is crucial for optimizing outcomes.

Standardizing procedures and optimizing protocols could reduce result variability and enhance the overall effectiveness of treatment.

The potential of these therapies extends beyond directly stimulating hair growth, also improving the tissue environment, making the scalp more receptive to other forms of treatment.

As research continues to explore and refine these techniques, the use of stromal vascular fraction cells emerges as a promising therapeutic option in hair loss surgery, offering new hope for those suffering from hair loss. However, it is essential to maintain a cautious and scientific approach, with particular attention to long-term result analysis, which appears to be one of the most promising aspects of these therapies.

5.7.4 Exosomes and Their Potential in Modulating the Hair Growth Cycle

In this context, exosomes, small extracellular vesicles, are emerging as a potential innovative therapy for patients with alopecia. In recent years, researchers have made significant progress in understanding how these tiny structures influence cellular interactions and signaling pathways involved in hair growth [12].

Exosomes derived from adipose-derived stem cells (ADSC-Exo) have proven particularly promising. These appear to counteract the negative effect of dihydrotestosterone on hair follicle growth. In vitro studies have shown that ADSC-exosomes promote the transition of hair follicles to the anagen (growth) phase by stimulating the proliferation of dermal papilla (DP) cells.

Exosomes operate through two primary mechanisms:

Paracrine signaling: Exosomes contain proteins, RNA, lipids, and other molecules that can be transferred to neighboring cells. This exchange influences a range of biological processes, including cell proliferation, differentiation, and apoptosis (programmed cell death).

Endocytosis: Cells take in exosomes through endocytosis, and their contents, including microRNAs, mRNAs, and proteins, are delivered into the cell, where they can modulate gene expression [3].

Exosome-based therapies could potentially reduce inflammation and oxidative stress in the scalp while improving the microenvironment for hair follicle growth. Additionally, because exosomes are naturally occurring vesicles, they are less likely to cause adverse immune reactions compared to stem cell therapies, making them a highly appealing alternative.

Exosome therapy is still in the early stages of clinical development, but preclinical studies and early-phase clinical trials have demonstrated positive results, especially in terms of increasing hair density and stimulating hair growth. However, further research, including large-scale clinical trials, is required to assess their long-term safety and efficacy.

5.8 In Conclusion

The growing interest in stem cell therapies, including those derived from adipose tissue, conditioned media, and exosomes, signals a promising future for the treatment of androgenetic alopecia and other forms of hair loss. While these therapies present exciting opportunities, their clinical application requires careful consideration of safety, regulatory challenges, and efficacy. Ongoing research is essential to optimize these treatments, standardize protocols, and explore their potential in combination with traditional therapies for enhanced results.

5.9 Sixth Paragraph: SVF/ADSC in Baldness Surgery

Fat tissue transplantation is a well-established clinical practice that has been in use for over 30 years, with its applications progressively expanding due to the increasing knowledge of the multiple functions of adipose tissue and its regenerative potential. However, the use of this approach in baldness surgery, aimed at harnessing the regenerative therapeutic action of adipose stem cells, has faced several practical challenges.

One initial difficulty relates to the limited surgical background of many professionals in the field of baldness surgery, who are often hesitant to include fat harvesting in their practice, being more accustomed to less complex techniques. Additionally, there was the challenge of developing a minimally traumatic, low-risk procedure—an essential requirement for a trichological patient—that would allow the effective use of adipose tissue in a compact structure like the scalp.

It is in this context that the guided SEFFI technique has made a significant impact, not only for its simplicity, which is well described in other parts of this book, but also for the opportunity it offers to use only the vascular stromal fraction (SVF). This is obtained by emulsifying and centrifuging microfragmented adipose tissue, providing a simple and safe clinical solution for scalp treatments. By avoiding complex methods such as enzymatic digestion or cell isolation in culture, the guided SEFFI technique simplifies and standardizes autologous tissue regeneration (ATR), ensuring greater safety.

SVF cells and their subpopulation of mesenchymal stem cells (ADSC), once isolated, have demonstrated the ability to promote and sustain the anagen phase of the hair cycle and exhibit anti-inflammatory and proangiogenic effects [8, 9]. These properties have been employed to maximize hair transplant results and potentially extend their duration over time.

Restoring the natural functions of the scalp is the key concept of this method, and this is why the use of the guided SEFFI technique can be particularly valuable in three areas of baldness surgery:

Before hair transplantation, to improve graft take and follicular regrowth.
In the correction of unsatisfactory outcomes from previous procedures, enhancing hair density and overall appearance.
In patients not suitable for new transplants, to improve scalp health and the condition of existing hair.

This cell-mediated therapeutic approach offers new opportunities to optimize hair transplant outcomes and contribute to greater patient satisfaction. However, it is essential to adhere to strict safety and procedural standards to ensure the effectiveness and safety of this innovative treatment.

5.9.1 Pre- and Posttreatment Instructions for Guided SEFFI Regenerative Therapy

Before a regenerative medicine procedure using the guided SEFFI technique, we share a series of pre- and posttreatment instructions with the patient, and we believe that great attention to these details has contributed to making our patients' experience pleasant and easily manageable during their everyday activities. The procedure is always explained in detail, and a comprehensive medical history and personal anamnesis are collected during the consultation.

It is required not to eat for at least 3 hours before the procedure, while clear liquids can be consumed up to 2 hours prior. Medications taken by the patient rarely pose a problem, and anticoagulants are generally suspended for a few days in agreement with the attending physician.

The procedure can be performed under local anesthesia, and to make the procedure even simpler and quicker, comfortable and easily removable clothing is recommended. To reduce the risk of infections, patients are advised to shower and thoroughly cleanse the scalp before the procedure.

After the guided SEFFI treatment, it is crucial to follow certain guidelines to ensure proper healing and optimize results:

Rest for a few hours, and to alleviate any mild soreness that may occur at the donor site, it is advised to take paracetamol or an NSAID.
Do not remove dressings at the donor site for the first 48 hours, unless otherwise instructed by the doctor.
Avoid shampooing and hair treatments during the first 2 days.
Perform regular check-ups as recommended by the doctor, and be mindful of any modest bleeding that may occur in the first 12 hours, which is considered normal.
Body temperature should be monitored on the evening of the procedure and the following morning, and in the rarest cases (only one in our clinical experience), broad-spectrum antibiotics may be prescribed for 5 days.

By following these simple instructions, optimal results can be achieved, and comfort during the recovery period can be maintained.

5.10 Seventh Paragraph: Guided SEFFI and Hair Transplantation

Regenerative medicine is opening up promising new perspectives in the treatment of androgenetic alopecia. Among the emerging innovations, the use of guided SEFFI is generating growing interest for its potential applications in hair transplants.

In the previous paragraphs, we examined the scientific foundation supporting the use of guided SEFFI, an innovative device that allows for the practical clinical application of adipose-derived stem cells.

Below, we will present the analysis of the results from three case samples, involving a total of 160 patients, in which the use of guided SEFFI was integrated into the hair transplant protocol.

5.10.1 Guided SEFFI in Preparation for Primary Hair Transplantation

In this section, we will examine how the preparation of donor tissue with guided SEFFI can improve the suitability and aesthetic outcome of primary hair transplants. This methodology supports the vitality of hair follicles, increasing their survival rate after transplantation.

In this group, we studied a sample of 67 patients. The average age of participants was 42 years. All patients were treated with a combination of SEFFI and hair transplantation (HT).

In 15 patients, or 23% of the total, a slight aesthetic improvement was observed after the SEFFI treatment. Additionally, 37 patients, accounting for 55% of the total, experienced earlier hair regrowth compared to the usual timeframes after the hair transplant, with visible results already seen by the third month post-surgery.

All patients are under long-term observation to assess the stability of the results compared to untreated patients.

These data suggest that the integration of SEFFI can provide visible benefits, both in terms of aesthetics and hair regrowth timing, improving the overall treatment outcomes.

Ultimately, measurable positive effects were observed in more than half of the patients. No complications were reported.

Through images of various treated cases, we will demonstrate how this method led to significantly superior results, both in terms of regrowth timing and the quality of transplanted hair.

5.10.1.1 Clinical Cases

Case 1: Guided SEFFI Treatment and HT for AGA Affecting the Frontal Lock and Mid-Scalp

Patient: Male, 53 years old.

Diagnosis: AGA (HAM V), located in the anterior and medial scalp regions.

Treatments:

2021: Guided SEFFI treatment

2023: Hair transplant (FUE) with 3050 follicular units (FUs)

Case Description:

The patient presents with advanced AGA (HAM V) and significant hair loss in the frontal and medial areas of the scalp. In 2021, the patient underwent a guided SEFFI treatment, which broke a therapeutic "plateau," resulting in an improvement in the caliber of native hair, contributing to greater density and creating a more receptive area for the transplant. Subsequently, in 2023, the patient underwent an FUE procedure, with the transplantation of 3050 follicular units, aimed at restoring hair in the areas affected by alopecia (Fig. 5.5).

Case 2: Guided SEFFI Treatment and HT for AGA Affecting the Frontal Lock and Mid-Scalp

Patient: Male, 49 years old.

Diagnosis: Type NORWOOD 4a AGA, located in the anterior region.

Treatments:

2020: Guided SEFFI treatment

2021: Hair transplant (FUE)—1956 follicular units (FUs)

Case Description**:**

The patient presents with grade 4 HAM AGA. In 2020, the patient underwent a guided SEFFI treatment in preparation for the FUE procedure, which was performed 4 months later. This treatment resulted in a synergistic improvement in

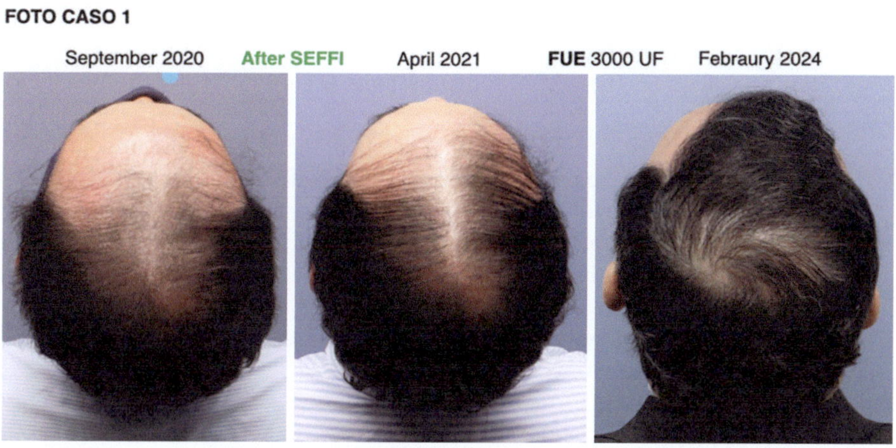

Fig. 5.5 Case 1 Clinical progression of hair restoration in a male patient (Case 1) treated with SEFFI and later FUE (3000 follicular units). Images from September 2020 to February 2024 show progressive improvement in hair density and coverage, highlighting the effectiveness of sequential regenerative and transplant therapies

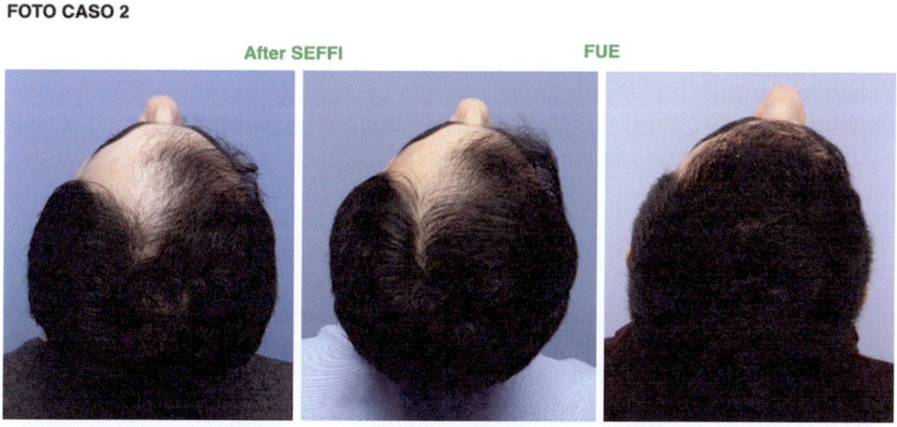

Fig. 5.6 Case 2 Hair restoration progression in a male patient (Case 2) following SEFFI and subsequent FUE treatment. The sequence of images illustrates a significant increase in hair density and crown coverage, highlighting the synergistic effect of regenerative and surgical therapies

the caliber of native hair, contributing to greater density in the anterior region. Subsequently, in 2021, the patient underwent an FUE procedure involving 1956 follicular units to thicken the anterior area (Fig. 5.6).

Case 3: Guided SEFFI Treatment and HT for AGA of the Anterior Area, Middle Scalp, and Crown
Patient: Male, 49 years old.
Diagnosis: HAM V type AGA.

Treatments:
2021: Guided SEFFI treatment
2021: Hair transplant (FUE)—3547 follicular units (FUs)
Case Description:
A 52-year-old patient with grade V HAM alopecia in the anterior and middle areas. Prior to the procedure, the patient underwent a guided SEFFI treatment, followed by an FUE procedure involving 3547 follicular units. Just 3 months post-surgery (see interim photo), notable hair regrowth is already visible, with further improvements observed in the following months (Fig. 5.7).

Case 4: Guided SEFFI Treatment and Hair Transplant (HT) for AGA Localized on the Vertex
Patient: Male, 57 years old.

Diagnosis: HAM VI type AGA, localized on the vertex.
Treatments:
2022: Guided SEFFI treatment
2023: Hair transplant (FUE)—2817 follicular units (FUs)
Case Description:
The patient, a 57-year-old man, presents with HAM VI type AGA, with hair loss predominantly in the crown area. The patient underwent a guided SEFFI procedure in 2022 to improve the quality of the scalp and prepare it for a subsequent hair transplant. In 2023, the patient underwent a hair transplant (FUE), with the transplantation of 2817 follicular units, resulting in significant hair regrowth and improved hair density in the treated area (Fig. 5.8).

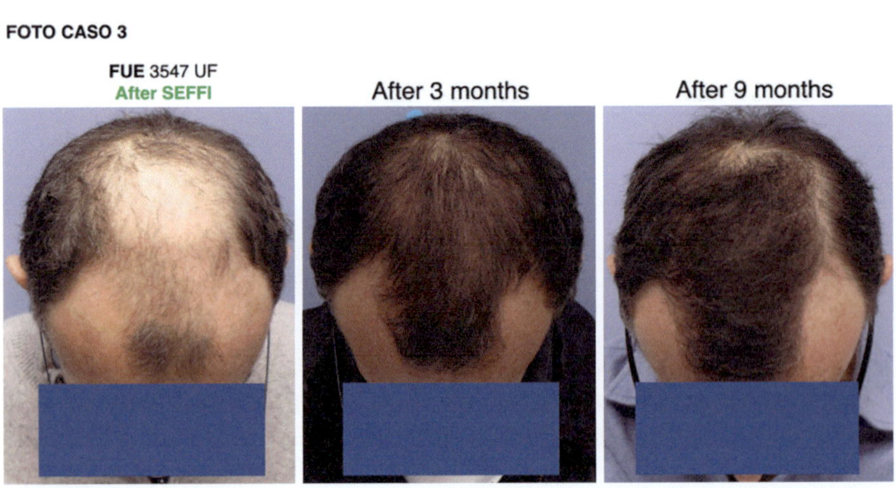

Fig. 5.7 Case 3 Hair restoration outcome in a male patient (Case 3) following SEFFI and FUE with 3547 follicular units. Clinical photographs show progressive improvement at 3 and 9 months posttreatment, demonstrating the effectiveness of combined therapies in restoring frontal hair density

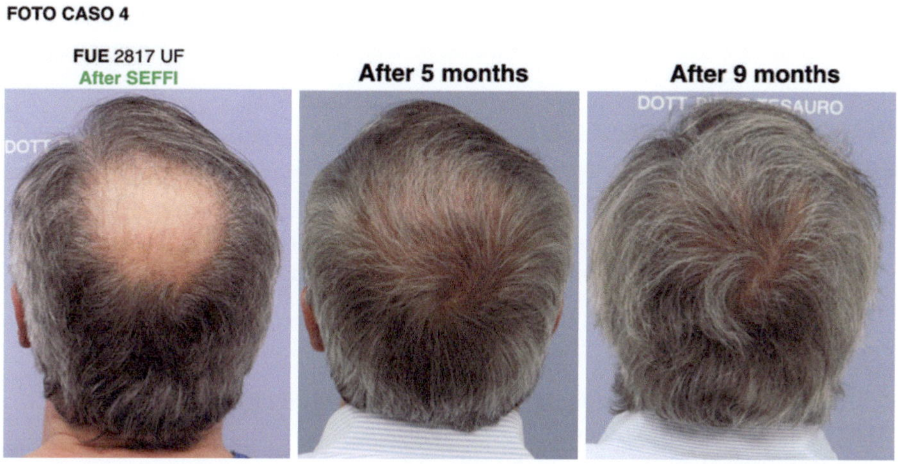

Fig. 5.8 Case 4 Hair regrowth progression in a male patient (Case 4) treated with SEFFI and FUE (2817 follicular units). Crown area shows marked improvement at 5 and 9 months, demonstrating the regenerative potential of combined therapeutic approaches in advanced androgenetic alopecia

1. Guided SEFFI in Preparation for a Secondary Hair Transplant.

In this second section, we will analyze the cases of patients with unsatisfactory results from previous hair transplants, who have benefitted significantly from the use of guided SEFFI. This technology has been successfully employed to revitalize previously transplanted miniaturized follicles and improve the quality of the recipient area before a corrective procedure, thus increasing the chances of achieving more satisfactory and long-lasting aesthetic results.

It is well-known that the recipient area of a patient who has already undergone a transplant can present particularly challenging characteristics, such as reduced vascularization and general tissue depletion. These factors can pose significant obstacles for a new transplant, negatively affecting follicle survival and overall hair density.

Our experience aims to demonstrate that a corrective transplant supported by regenerative treatment can offer a higher likelihood of success, even in the most difficult and complex cases.

This group includes a sample of 86 patients with an average age of 47, who sought to improve the results of previous hair transplants with a second procedure. On average, the time elapsed since the previous procedure was 5 years.

The results obtained were noteworthy: in 18 patients, accounting for 21% of the total, an appreciable improvement in hair caliber was observed before proceeding with the second transplant. Additionally, 26 patients, representing 30% of the cases, experienced early hair regrowth as early as 3 months after the procedure. A particularly significant finding was that in 40 patients, approximately half of the sample, the implantation phase was notably easier and more straightforward, very similar to a primary hair transplant, facilitating the execution of the procedure and likely contributing to a faster recovery.

The following clinical cases confirm that the adopted therapeutic approach can offer significant advantages, improving both scalp quality and the management of the procedure itself.

5.10.1.2 Clinical Cases

Case 5: Guided SEFFI Treatment and Hair Transplant (HT) Following Unsatisfactory Results from Two Previous Hair Transplant Procedures

Patient: Male, 40 years old.

Diagnosis: AGA (HAM IV).

Previous Treatments:

2013 and 2016: Two strip technique hair transplants with unsatisfactory results

2020: Guided SEFFI treatment

2022: Hair transplant (FUE)—2236 follicular units (FUs)

Case Description:

The patient, a 40-year-old man, previously underwent two hair transplants (FUE) in 2013 and 2016, but both procedures resulted in unsatisfactory outcomes. In 2020, the patient underwent a guided SEFFI procedure. This treatment showed noticeable improvements within 3 months, with an increase in the caliber of both the transplanted and native hair. Subsequently, in 2022, the patient underwent a new FUE hair transplant involving 2236 follicular units, achieving a significant improvement in coverage and density in the treated areas (Fig. 5.9).

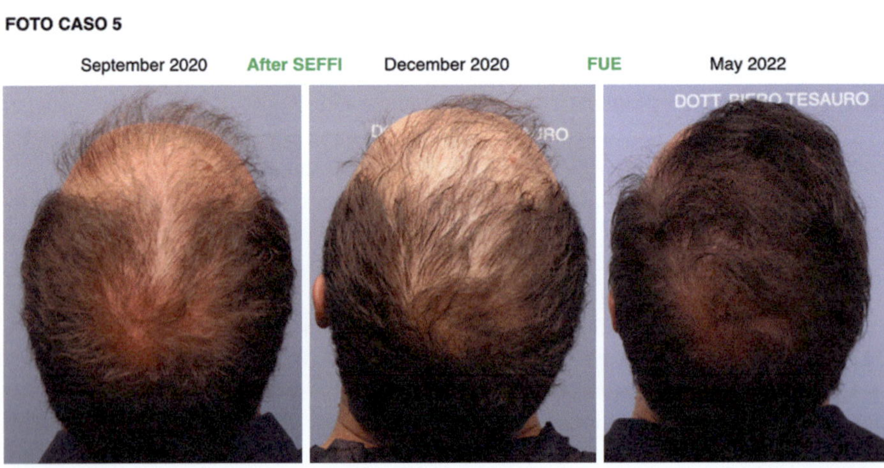

Fig. 5.9 Case 5 Hair restoration progression in a male patient (Case 5) treated with SEFFI followed by FUE. Clinical photographs from September 2020 to May 2022 show substantial improvement in scalp coverage and hair density, particularly in the crown area, highlighting the long-term efficacy of combined regenerative and transplant techniques

Case 6: Guided SEFFI Treatment and Hair Transplant (HT) Following Results of a Previous Hair Transplant Procedure
Patient: Male, 35 years old.
Diagnosis: Diffuse thinning—AGA (HAM III).
Previous Treatments:
2016: FUE—1715 follicular units (FUs)
2020: Guided SEFFI treatment
2022: FUE—2400 follicular units (FUs)
Case Description:
The patient presented with AGA (HAM III). In 2016, he underwent an initial FUE hair transplant of 1715 follicular units to treat the frontal area. In 2020, the patient underwent guided SEFFI treatment to improve the results and prepare the scalp for a second procedure. After the treatment, there was noticeable improvement in both the density and quality of the native and transplanted hair. Subsequently, a second transplant of 2700 follicular units was performed. Thanks to the guided SEFFI treatment, the scalp became soft and easily treatable, facilitating the second procedure (Fig. 5.10).

Case 7: Guided SEFFI Treatment in a Patient Who Underwent an FUE Procedure, in Preparation for the Second and Third Procedures
Patient: Male, 53 years old.

Diagnosis: AGA HAM III.
Previous Treatments:
2018: FUE—1864 follicular units (FUs)
2020: Guided SEFFI treatment
2021: FUE—1508 follicular units (FUs)
2023: FUE—1000 follicular units (FUs)

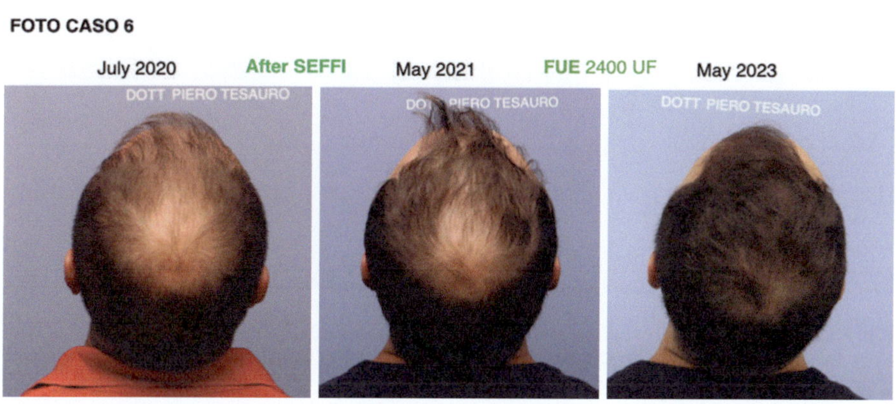

Fig. 5.10 Case 6 Hair restoration progression in a male patient (Case 6) following SEFFI and FUE with 2400 follicular units. Over a period of 3 years, clinical images demonstrate substantial improvement in crown density and coverage, underscoring the effectiveness of sequential regenerative and surgical therapies

Case Description:

This patient presented with AGA and a long therapeutic history. In 2018, he underwent his first FUE hair transplant with 1864 follicular units. Two years later, to optimize the results and prepare for subsequent procedures, he opted for a guided SEFFI treatment to improve the condition of his scalp. The combination of guided SEFFI and FUE again led to an excellent outcome (Fig. 5.11).

2. Guided SEFFI in Patients with Secondary Outcomes Not Undergoing Further Surgery.

In this final section, we will explore the role of guided SEFFI treatment as an opportunity for patients who are not candidates for further hair transplants or who choose not to undergo additional procedures. "Non-reoperable patients" refer to those with a depleted donor area following previous procedures or a low hair density per square centimeter, making further corrective transplants unfeasible. This category also includes patients who, by choice, do not wish to have another hair transplant and seek alternative solutions.

In these cases, guided SEFFI treatment can be integrated into the therapeutic plan to promote hair regrowth and increase the thickness of existing hair, utilizing the mechanisms already described in the previous sections.

We analyzed a sample of seven patients. Of these, 4 patients, representing 57% of the total, showed a noticeable improvement in hair caliber.

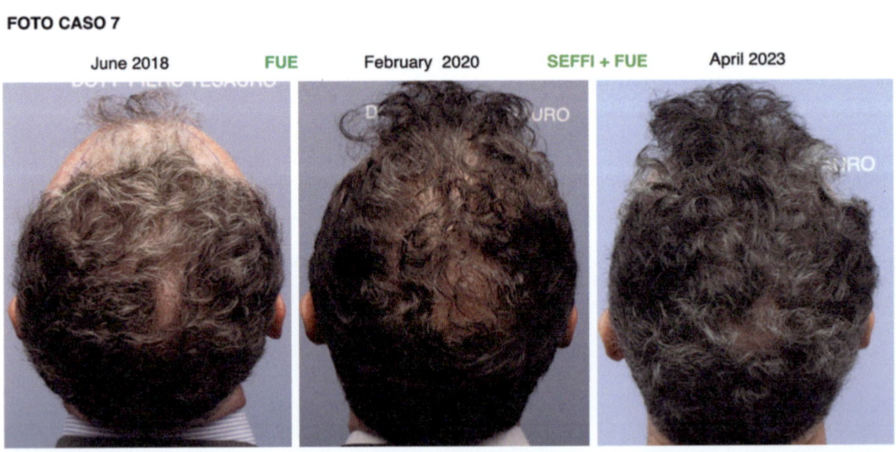

Fig. 5.11 Case 7 Long-term hair restoration in a male patient (Case 7) with curly hair, treated with FUE followed by combined SEFFI and FUE. Images from 2018 to 2023 show progressive improvement in crown density, demonstrating the sustained efficacy of integrated regenerative and transplant techniques over time

5.10.1.3 Clinical Cases

Case 8: Guided SEFFI Treatment in a Non-reoperable Patient

Patient: Male, 56 years old.

Diagnosis: AGA (HAM VI).

Previous Treatments**:**

Two Strip procedures performed in 2013.

Case Description:

A 56-year-old patient with two previous hair transplants who is not a candidate for further procedures due to a lack of follicular availability in the donor area. In 2020, the patient underwent a guided SEFFI procedure. Six months later, there was a noticeable increase in hair density, both in transplanted and native hair, with a reduction in transparent areas (Fig. 5.12).

Case 9: Guided SEFFI Treatment Following Two Hair Transplants to Improve Results

Patient: Male, 53 years old.

Diagnosis: AGA HAM II.

Previous Treatments:

2014: Strip—1826 follicular units (UF)

2017: FUE—1695 follicular units (UF)

2020: Guided SEFFI procedur

Case Description**:**

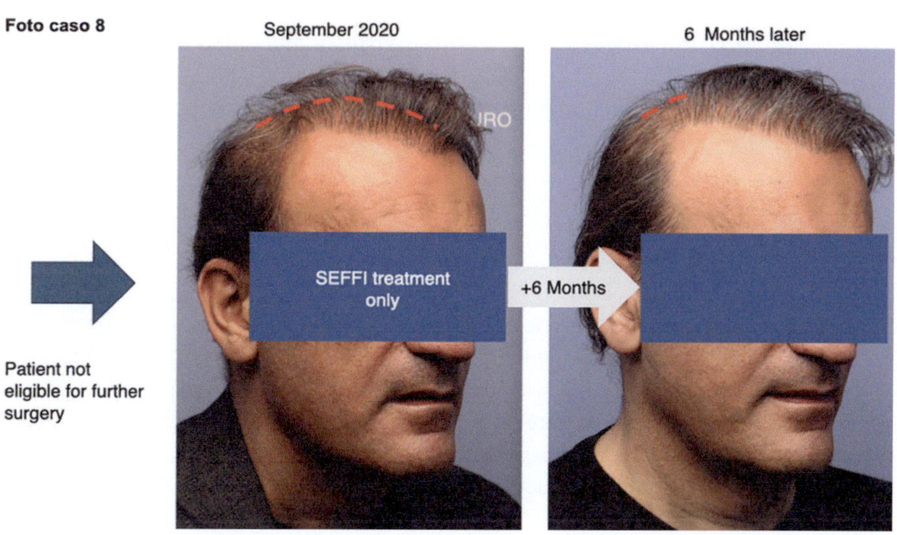

Fig. 5.12 Case 8 Hair density improvement in a non-reoperable male patient (Case 8) treated exclusively with SEFFI. Comparative photographs before and 6 months after treatment show modest yet measurable enhancement of frontal scalp coverage, supporting the role of regenerative therapy in cases not eligible for surgical hair restoration

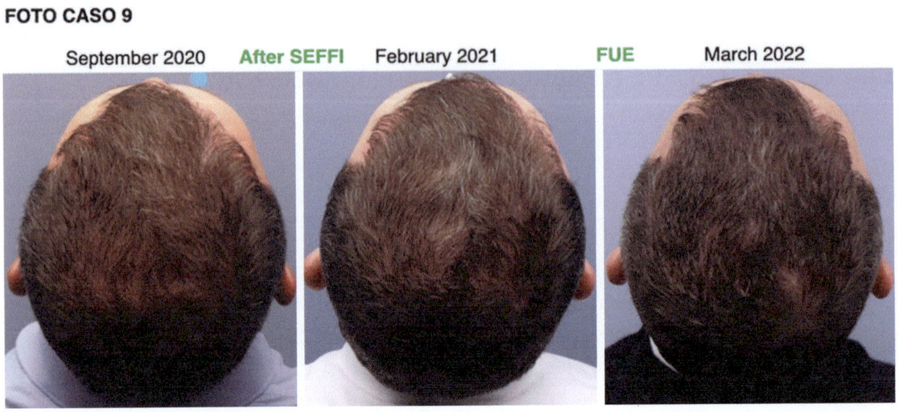

Fig. 5.13 Case 9 Hair restoration in a male patient (Case 9) treated with SEFFI followed by FUE. Photographs taken from September 2020 to March 2022 show gradual and consistent improvement in crown hair density, confirming the synergistic effect of regenerative and transplant techniques

The patient presented with thinning hair due to AGA grade II according to the Hamilton scale. In 2014, he underwent a hair transplant using the strip technique, with generally positive results, although some small areas of postoperative alopecia appeared, which subsequently resolved. After 2 years of substantial stability, the patient noticed a gradual increase in hair loss and decided to undergo a second transplant using the FUE technique, with 1965 follicular units. The result was satisfactory, and the patient maintained good stability for several years. In 2020, the patient sought a third transplant and opted for guided SEFFI treatment preoperatively. Following the treatment, the condition significantly improved, and the patient no longer felt the need for further surgical intervention (Fig. 5.13).

5.11 Conclusion

History in medicine often invites reflection on the transition from traditional approaches to innovative ones, which frequently occur as a result of groundbreaking discoveries. This was the case with organ transplants, made possible through the study of immunosuppression, targeted chemotherapy, which is increasingly effective because it is tailored to the genetic characteristics of cancer, and minimally invasive surgery, which has reduced risks and improved recovery times for many procedures.

Today, we are accumulating significant clinical experience that is gradually expanding the role of regenerative medicine in the treatment of baldness, and our experience with stem cell use has harnessed a simple yet innovative tool, guided SEFFI, to achieve safer and more durable results.

We are still in a phase where we are seeking the optimal window of opportunity for this therapy. We hope that more precise studies of the biological mechanisms underlying our successes will enable even more targeted applications in the future.

In the Ebers Papyrus, the Egyptian physician Hakiem el Demag, over 4000 years ago, suggested applying compresses made from the fat of various animal species to the heads of patients suffering from alopecia, leaving them for 4 days. Today, we can hypothesize that he had an incredible intuition. While he certainly knew nothing of exosomes or mRNA, it is fair to believe that if he subjected his patients to these small tortures, he must have had some positive outcomes.

The dialogue between intuition and scientific knowledge is a crucial part of scientific development. Many discoveries arise from observations and intuitions, even before a complete understanding is achieved. Our experience with guided SEFFI in hair loss surgery demonstrates this concept: despite not fully understanding the mechanisms behind the interaction between VSF and HF, the results obtained and the energy generated from positive feedback drive us to continue, knowing that there is much more to discover; continuing on this path could prove to be the starting point for significant future progress.

References

1. Schneider MR, Schmidt-Ullrich R, Paus R. The hair follicle as a dynamic miniorgan. Curr Biol. 2009;19(3):R132–42.
2. Stenn KS, Paus R. Controls of hair follicle cycling. Physiol Rev. 2001;81:449–94.
3. Rishikaysh P, Dev K, Diaz D, Qureshi WM, Filip S, Mokry J. Signaling involved in hair follicle morphogenesis and development. Int J Mol Sci. 2014;15(1):1647–70.
4. Treb RM. Molecular mechanisms of androgenetic alopecia. Exp Gerontol. 2002;37(8–9):981–90.
5. Sinclair RD. Male androgenetic alopecia. JMHG. 2004;1(4):319–27.
6. Messenger AG, Rundegren J. Minoxidil: mechanisms of action on hair growth. Br J Dermatol. 2004;150(2):186–94.
7. Mella JM, Perret MC, Manzotti M, Catalano HN, Guyatt G. Efficacy and safety of finasteride therapy for androgenetic alopecia: a systematic review. Arch Dermatol. 2010;146(10):1141.
8. Piccinno MS, Petrachi T, Pignatti M, Murgia A, Grisendi G, Candini O, Resca E, Bergamini V, Ganzerli F, Portone A, Mastrolia I, Chiavelli C, Castelli I, Bernabei D, Tagliazucchi M, Bonetti E, Lolli F, De Santis G, Dominici M, Veronesi E. Human adipose mesenchymal stromal/stem cells improve fat transplantation performance. Cells. 2022;11(18):2799.
9. Krefft-Trzciniecka K, Piętowska Z, Nowicka D, Szepietowski JC. Human stem cell use in androgenetic alopecia: a systematic review. Cells. 2023;12(6):951.
10. Ji S, Zhu Z, Sun X, Fu X. Functional hair follicle regeneration: an updated review. Signal Transduct Target Ther. 2021;6(1):66.
11. Rahmani W, Abbasi S, Hagner A, Raharjo E, Kumar R, Hotta A, Magness S, Mezge Biernaskie J. Hair follicle dermal stem cells regenerate the dermal sheath, repopulate the dermal papilla, and modulate hair type. Dev Cell. 2014;31(5):543–58.
12. Tang X, Cao C, Liang Y, Han L, Tu B, Yu M, Wan M. Adipose-derived stem cell exosomes antagonize the inhibitory effect of dihydrotestosterone on hair follicle growth by activating Wnt/β-catenin pathway. Stem Cells Int. 2023;2023(1):5548112.

13. Villani A, Ferrillo M, Fabbrocini G, Ocampo-Garza SS, Scalvenzi M, Ruggiero A. Hair aging and hair disorders in elderly patients. Int J Trichology. 2022;14(6):191–6.
14. Tobin DJ, Magerl M, Gunin A, Handijski B, Paus R. Plasticity and Cytokinetic dynamics of the hair follicle mesenchyme: implications for hair growth control. J Invest Dermatol. 2003;120(6):895–904.
15. Jimenez F, Higgins C, editors. Hair follicle regeneration. Humana Press; 2022.

Chapter 6
Autologous Regenerative Therapy and Bio-photostimulation: A Surprising Combination

The **TRICHOBIOLIGHT® protocol** complements ART (Autologous Regenerative Therapy) through a structured regimen of **mesotherapy**, delivering **hyaluronic acid, growth factors, amino acids, and antioxidants** directly into the scalp. This is immediately followed by **red LED photobiostimulation (630 nm)**, enhancing cellular ATP production and increasing tissue receptivity to therapeutic agents. Clinical studies demonstrate that this **synergistic approach** significantly improves **hair thickness, density, and follicular stability**, with patients reporting high satisfaction rates and minimal side effects.

By targeting **both the follicular microenvironment and systemic regenerative pathways**, this multimodal therapy offers a **biologically sustainable alternative** to conventional treatments, positioning **stem cell-based regenerative strategies** at the forefront of **trichological medicine**.

6.1 ART and Photobiostimulation: A New Combined Protocol for Hair Loss

Nowadays hair is associated with an identity function, and its loss can cause, from its first manifestations, a serious discomfort for patients, causing a worsening of the quality of life with direct consequences on the overall well-being.
Hair loss can be associated to numerous causes depending on the problem observed or comorbidities, such as in acute telogen or some forms of alopecia areata, scarring alopecia, or in various autoimmune diseases (lichen, lupus) or iatrogenic lesions.

In most cases, however, hair loss recognizes a genetic origin, correlated with androgen hormone production (AGA), often aggravated by numerous incidences such as digestive, nervous, physical, or psychophysical wasting.

AGA affects 70% of men, who may show the first signs of the condition as early as late adolescence, and 40% of women generally affected from the third, fourth

D. Piccolo et al., *A.R.T. Autologous Regenerative Therapy in Aesthetic Medicine*, https://doi.org/10.1007/978-3-032-05835-5_6

decade onward. Men typically show the classic M-shaped pattern with recession of the hairline at the temples and hair loss at the apex, while women more frequently show diffuse thinning over the entire upper part of the scalp [1–3].

Dermoscopically, hair miniaturization, a reduction in the number of hair shafts per pilosebaceous unit (a single hair instead of 2–3) and numerous yellow dots are observed: evidence, in this pathology, of a "stuck" follicle in kenogen that fails to express a new growth phase. This examination allows early recognition of early signs of AGA, which become increasingly evident as the years go by and the pathology progresses. Patients almost invariably report a loss of mass and a progressive increase in transparencies [4, 5].

Androgenetic alopecia, recently described as a disease with autosomal dominant polygenic inheritance, is strongly associated with testosterone/dihydrotestosterone imbalance and marked receptor sensitivity to DHT both of which cause shortening of the anagen phase and early entry I follicle into telogen. Acceleration of the follicular cycle generates functional incapacity and subsequent miniaturization [6, 7].

Several therapeutic approaches to AGA of pharmacological, physical, and cosmeceutical types, prescribed to patients individually or in combination, have been tried in recent decades. The literature reports varying success rates, and the outcome depends mainly on the earliness of therapy and disease progression.

In this study, we decided to test a new type of outpatient protocol for the treatment of AGA by structuring a therapeutic process with three innovative methods placed in synergy with each other, each capable of enhancing the action of the previous one.

The first step of our protocol is based on the use of an adipose stem cell extract enriched with vasculostromal cells (SVF). For adipose tissue harvesting, we used a specific medical device manufactured by SEFFILINE S.r.l. Bologna—Italy [13, 16–25, 27, 28].

The word SEFFI stands for Superficial Enhanced Fluid Fat Injection [13, 16–18, 20–25, 27, 28].

For the second step, we chose to subject patients to a sequence of mesotherapy and photobiostimulation.

Mesotherapy, "therapy in the middle layer of the skin," is based on multiple superficial dermal microinjections and has shown significant potential in the treatment of AGA. In this protocol, we included a selection of natural substances [14] in the therapeutic cocktail, excluding pharmacological active ingredients and choosing the 1-month interval as the right compromise between efficacy and compliance [8–12].

Photobiostimulation with red LED light at 630 nm proved to be very effective in activating mitochondria and the ATP cycle, thus leading to increased receptivity of cells to nearby substances.

This second step represents the already structured and registered protocol, previously named TRICHOBIOLIGHT®, of which consistent results have already been published. This combined protocol, ART and photobiostimulation, was put in place with the aim of testing whether the results previously obtained with the latter alone could be further improved.

Specifically, we hypothesized that photobiostimulation with a 630 nm LED light would increase the receptivity of cells to nearby substances (in this case, those delivered by mesotherapy) and would promote the development of mesenchymal stem cells transplanted around hair follicles [15].

In this retrospective study, we present the results obtained in almost 2 years of experimentation.

6.2 Materials

We selected 40 Caucasian patients, 12 females and 28 males, with a mean age of 36.4 years (range 16–67) with a clinical diagnosis of mild to moderate AGA, grades II to V according to the Norwood-Hamilton classification. The patients were treated privately at our four "Novea Skin Centers," located in L'Aquila, Avezzano, Pescara, and Catania (Italy), from June 2020 to August 2022.
All patients signed informed consent.

Inclusion criteria in the protocol were clinical and trichoscopic diagnosis of AGA and suspension of other trichological therapies for at least 6 months while diagnosis of other scalp disorders and unavailability of complete adherence to the treatment protocol were the exclusion criteria.

The clinical history was followed by a detailed trichological history that included the age of onset of AGA, clinical history of the disease, and previously prescribed therapies.

Before starting the combined ART + photobiostimulation protocol, a global scalp photograph, pull test, trichogram, and trichoscopy were performed on all patients.

Trichoscopy allowed assessment of hair caliber and number by highlighting the percentage of anisotrichia and differences in number of hair shafts per follicular unit between the areas affected by the disease and the comparison occipital region. These data were used both to identify disease severity and to assess therapeutic response. The first step of the protocol to which the patients underwent was ART (Autologous Regenerative Treatment) with SVF and ADSC using the guided SEFFI device. (Fig. 6.1).

The procedure involves, once the harvest area is established (in most cases the abdomen or flanks), preparation of the surgical field, which is followed by local

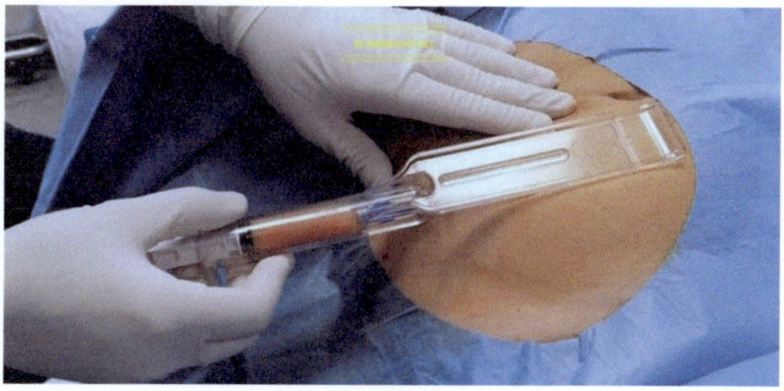

Fig. 6.1 Tissue grafting procedure using the guided SEFFI technique. A medical professional extracts micrografts with a specialized needle while the patient is draped in sterile blue surgical cloth, demonstrating standard operative conditions

anesthesia of the entry point and small incision at the margin of the selected area. Then, with the same instrument, both the infiltration of the anesthetic solution and, after a 10-minute pause necessary for vasoconstriction, the aspiration of the fatty tissue is performed. This instrument consists of a syringe with an automatic plunger lock on which a 2 mm.Ø cannula with 0.8 mm.Ø. microholes is screwed, inserted inside a special guide. The instrument ensures that tunneling is performed in the superficial subcutaneous tissue, next to the dermis, thus harvesting the ideal tissue that is richest in SVF and ADSCs, offering the best quantity and quality of cells [16–28].

Once collected, the adipose tissue, already microfragmented, is gently washed and then emulsified (A–B) by 20 passes between two syringes joined by a connector. The resulting product is then transferred into tubes and centrifuged for 10 minutes at 3500 rpm (Fig. 6.2).

At the end of the centrifugation, the infranatant (at the bottom of the tubes) is separated from the tissue (deposited at the top) and can be easily aspirated from the bottom of the tubes by means of a long needle; it is then ready so that it can be injected into the dermis of the scalp through a 4-mm 27G needle, easily and painlessly (Fig. 6.3) [22–24].

One month after a single session of ART (Fig. 6.4) begins the second phase of our protocol which consists of a combined therapeutic approach of mesotherapy and photobiostimulation called TRICHOBIOLIGHT® (trademark registered with the EUIPO -ID No. 2020/098-already available online at http://www.euipo.europa.Eu/).

Mesotherapy was performed using a 1.5-mL syringe equipped with a 30G, 4-mm needle. A total of 3 mL of a hyaluronic acid–based solution was injected, containing polydeoxyribonucleotides (5.625 mg/3 mL), seven vitamins, fibroblast growth factors, polypeptides, sodium DNA, 16 amino acids, eight minerals, Ginkgo biloba extracts, linoleic acid, oleic acid, quercetin, kaempferol, chlorogenic acid, gallic acid, and water. (Fig. 6.5) [21, 25, 27, 28].

The mesotherapy session is immediately followed by a photobiostimulation session with an E-Light technology scalp conformal lamp lasting 15 min. (Figs. 6.6 and 6.7).

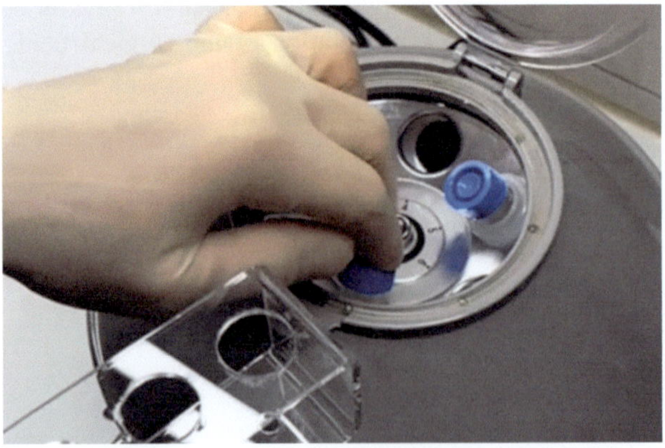

Fig. 6.2 Placement of a blue-capped vial into a centrifuge rotor by a gloved hand

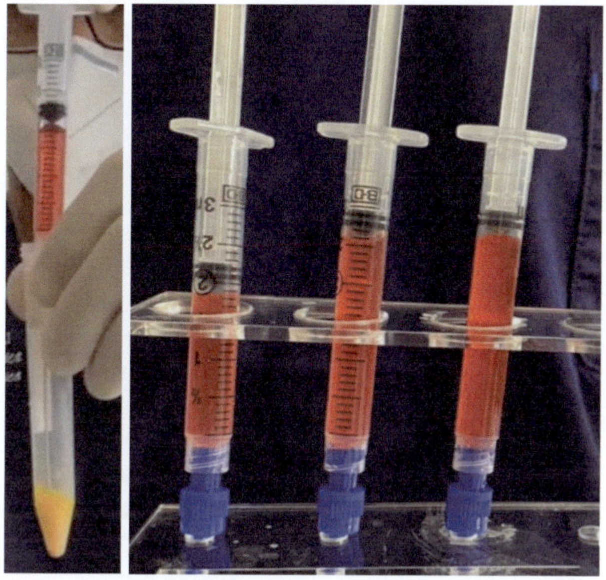

Fig. 6.3 Laboratory preparation of the infranatant portion following centrifugation. The image shows separated layers in a test tube and in syringes secured in a holder, with adipocyte-rich fractions visible. Blue syringe caps indicate readiness for further processing or clinical application

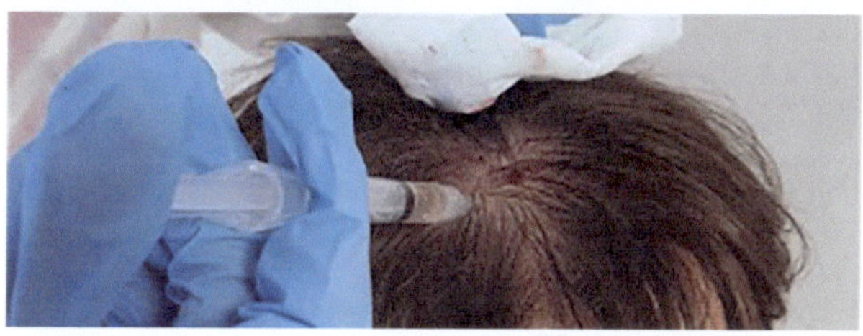

Fig. 6.4 Scalp injection of infranatant portion

Patients underwent six sessions of the protocol, each monthly.

Follow-up visits were made every 4 months, and one or two photobiostimulation maintenance sessions per year were prescribed, depending on the severity of the disease and the results obtained (evaluation of these protocols is being optimized). In addition, we asked each patient to complete a satisfaction questionnaire at the end of treatment.

"Hyaluronic Acid Plydeoxyribonucleotides /5,625mg/3mL)"		
Vit E Vit B12 Vit B9 Vit B3 Vit B5 Vit A	Fibroblasts GF + Polipeptides + NaDNA	Water
Quercetin Kaempferol	Arginine Lysine Alanine Methionine Aspartic acid Phenylalanine Proline. Glutamic acid Glycine Serine Histidine Threonine Tyrosine Hydrossyproline Isoleucine Leucine Valine	Silica Iron Potassium Magnesium Manganese Sodium Phosphorus Selenium
Bliobalide Ginkgolide ABCM	Chlorogenic Acid Gallic Acid	Lynoleum Acid Oleci Acid

Fig. 6.5 Composition table of a product containing hyaluronic acid and polydeoxyribonucleotides (5.625 mg/3 mL). The table categorizes active ingredients into vitamins, growth factors, amino acids, minerals, plant-based compounds, and fatty acids, highlighting the multifactorial formulation intended for regenerative or aesthetic applications

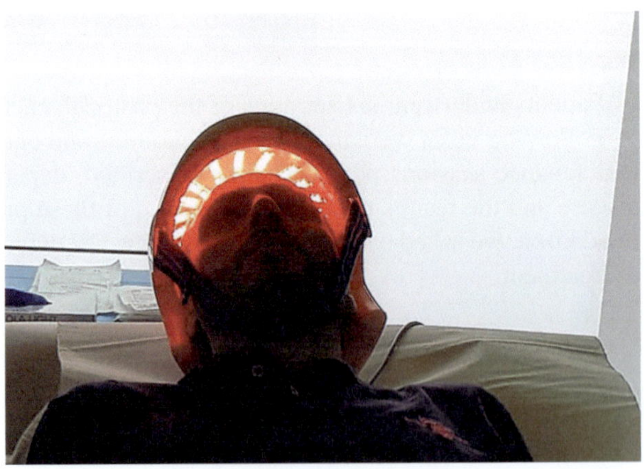

Fig. 6.6 Light therapy session using a red-light emitting helmet. The patient lies in a clinical setting during the procedure, which is designed to stimulate scalp or skin regeneration

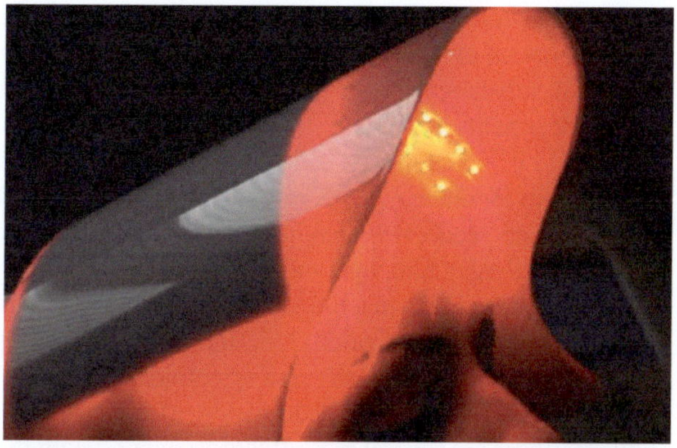

Fig. 6.7 Red light therapy session targeting the forehead using an LED device

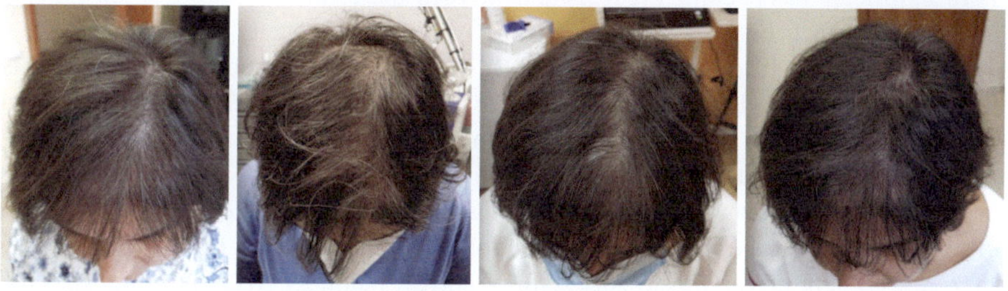

Fig. 6.8 Sequential top-view images documenting hair density improvement over time. The clinical setting highlights the context of therapeutic intervention for hair regrowth

6.3 Results

All patients responded positively to the treatment; an excellent result was achieved in 82.5% of patients (33/40) while good result was observed in 7 patients (17.5%). Indeed, the collected data demonstrated the excellent efficacy of our protocol in the treatment of baldness, showing an early clinical response even after only 4 months of therapy, with results that were maintained in the following 12–18 months [21, 25, 26, 28].

At follow-up visits, instrumental examinations provided confirmation of the clinical evidence. Trichograms showed an increase in the percentage of hair in anagen while trichoscopy showed a reduction in anisotrichia and a 20% increase in hair shaft diameter (Figs. 6.1 and 6.8).

Common side effects showed mild to moderate pain after the mesotherapy session and/or an itching or burning sensation that always resolved spontaneously within 24 hours [25, 26, 28].

At the end-of-treatment satisfaction questionnaire, 97.5 percent (39/40) of patients reported high satisfaction with the treatment, related to improved perception of their image due to a visible increase in hair thickness and shine, associated with a decrease in hair loss. In recent decades, more and more studies have been published with the aim of finding an effective therapy for AGA, especially since baldness has a significant psychological impact on the quality of life of affected patients.

The most common therapies proposed to patients are natural therapies, of which a wide array of stimulating lotions and dietary supplements with special reference to natural substances with antiandrogenic and antioxidant effects and pharmacological treatments are part.

It should be noted that so-called 'natural' therapies based on herbal extracts such as saw palmetto (*Serenoa repens*), rosemary (*Rosmarinus officinalis*), licorice (*Glycyrrhiza glabra*), grape seed (*Vitis vinifera*), pumpkin seed (*Cucurbita pepo*), and green tea (*Camellia sinensis*), although widely used, often yield inconsistent clinical results [1].

FDA-approved pharmacological treatments for the medical treatment of AGA are based on formulations of minoxidil, finasteride, and dutasteride: These substances can be administered both topically and systemically, and due to their demonstrated action on the androgen, hormone cascade often offers significant and stable results over time. However, there are several side effects associated with the use of these drugs, and even more significant has become the resistance to their use related to fear over the appearance of side effects on the sexual sphere or psychological sphere [8, 11].

We would like to emphasize how both of the described therapeutic proposals delegate to the properly educated patient the adherence to the treatment plan both in terms of adherence to the therapeutic scheme and to the mode of use/application of the prescribed products.

In addition, it is also worth noting that, in addition to or instead of therapies, there is an exponential increase in self-transplantation surgeries (FUE or FUS), which today offer an excellent possibility to successfully correct baldness. Without going into the details of this method, it is clear that the limitations of the surgical approach are inextricably linked to the capacity of the donor area and the aggressiveness of the baldness. When surgery is performed on a still evolving form of AGA, the real success of this approach should be understood as the stable maintenance of the result without resorting to early supplementation. It is also important to highlight that, alongside medical therapies, the number of self-transplantation procedures (FUE and FUS) has increased exponentially in recent years, providing an effective option for restoring hair density. Although the details of these techniques are beyond the scope of this work, it should be acknowledged that the success of any surgical approach is intrinsically limited by the donor area capacity and by the severity and progression of androgenetic alopecia. When surgery is performed in a still-evolving form of AGA, long-term success depends primarily on the stable maintenance of the result without the need for early adjunctive treatments. This once again underscores the importance of achieving an adequate therapeutic balance before surgery whenever possible. Such stability, when obtained, represents the most favorable condition for achieving an optimal long-term outcome after autotransplantation.

This therapeutic combination, an alternative to conventional therapies, allows the patient to be accompanied, step by step, until the first therapeutic successes without delegation. At the same time, it avoids the use of drugs and promotes a good acceptance of therapy by creating the right conditions for an excellent therapeutic adherence further fostered by the positivity of the results obtained by motivating patients to keep their monthly appointments on time.

We are aware in our clinical research activities that these protocols act on molecular mechanisms about which we still know relatively little. Work published in recent years has shown that the hair follicle and dermal white adipose tissue (dWAT) constitute two compartments that, since late embryogenesis, are physiologically coupled. This epithelium/mesenchyme interaction is based on well-defined pathways traversed by paracrine signals and is essential for the tissue to be healthy. (Ramos).

In AGA, this balance is altered and miniaturized follicles are characterized by a decrease in the number of cells within the dermal papilla (DP), but this condition is not matched by a reduction in the number of hair follicle stem cells (HFSCs) located in the mesenchyme at the base of the HF.

These observations have led researchers to hypothesize two possible causes of the pathology-related miniaturization, the first related to an intrinsic defect of HFSCs located in the bulge region that makes them unable to convert into progenitor cells, while the second mechanism proposes that there is an altered or weakened signal from the DP to the HFSC bulge.

Stem cell inoculation in sufficient concentration aims to unlock both of the above mechanisms by trying to reverse the pathogenesis of hair loss by reactivating dermal papilla cells or activating HFSCs in the bulge. This action would be carried out through paracrine signals, specifically cytokines and growth factors such as platelet-derived growth factor (PDGF), vascular endothelial growth factor (VEGF), and insulin-like growth factors (IGFBP-1 and IGFBP-2 binding proteins), which when secreted go on to stimulate hair follicle stem cells (HFSCs) and niche cells.

The efficacy of the proposed therapy is based on the ability of the implanted cells (ADScs and VSCs) to induce regeneration of contiguous follicular structures, positively supported by the nutrients and energy provided by mesotherapy and photobiostimulation sessions.

In essence, this heterogeneous population of cells is capable of transforming the dermal macroenvironment whose importance has been repeatedly emphasized.

The photobiostimulation system had already proven extremely effective in AGA without additional treatments. Mesotherapy, which is its first phase, allows direct delivery of the active ingredients into the scalp at an exact depth, to the effect of which is added the local release of cytokines and growth factors induced by the microtrauma induced by the sequential injections. The benefits of this method, which offers the possibility of administering numerous substances at different times and concentrations, are widely described in the literature [21, 25, 27, 28].

Compared with other proposed therapies, the combination of the ART and photobiostimulation protocol is distinguished by a number of advantages:

- The use of autologous regenerative treatment increases the probability of success of subsequent stimulation steps.
- Mesotherapy guarantees precise administration on the affected area, unlike topical administrations of other active ingredients, which must be performed on the entire scalp with a whole series of more or less complex precautions for the patient.
- The mesotherapy method has the advantage of reducing potential adverse reactions or intolerance toward the active ingredient and the additional use of photobiostimulation by means of a red LED light at 630 nm, which has already been shown to be very effective in increasing the receptivity of cells to substances (in this case those carried by mesotherapy) in the vicinity [27, 28].

We decided to use the guided SEFFI kit because of both its efficacy and the peculiar characteristics of this device, made with the aim of making the whole procedure safe, reproducible, and simple for both the experienced physician and the physician with no previous expertise in liposuction, offering to the latter, a very fast and effective learning curve.

This method also makes it possible to use the regenerative potential of SVF and ADSCs while avoiding the problems inherent in the difficult and painful injection of microfragmented adipose tissue into the scalp, which in fact does not lend itself to being infiltrated with cannulas or needles except with great difficulty [13, 16–20, 22–24, 28].

Hair loss can lead to moderate to severe psychological distress, and androgenetic alopecia represents the most common condition for which countless treatments, not always effective, are now offered to patients. These often fail to take into account some basic needs of patients such as reliability, which is essential in order not to waste the initial enthusiasm with which patients approach treatment, support during the course of treatment, and the shared request to avoid side effects [27, 28].

As a result of years of study and research, our protocol, which proposes the combination of three well-known therapies such as autologous regenerative transplantation, mesotherapy, and photobiostimulation, has proven effective in all cases, treated regardless of their severity [21, 25–28].

This is due to the fact that both ART and TRICHOBIOLIGHT act on existing structures that are still able to regenerate when properly stimulated, with a significant reduction in side effects.

This combined ART and photobiostimulation protocol has resulted in considerable patient satisfaction, often surprised by having achieved unhoped-for results such as the recovery of hair that they considered lost moreover in a much shorter time than expected. This has generated an increased sense of self-esteem and confidence that, in our opinion, other techniques cannot guarantee [27, 28].

Among the advantages of this method, we would also like to highlight the punctuality in the application of therapy, guaranteed by the fact that it cannot be delegated to the patient but performed exclusively by the doctor throughout the duration of treatment. The absence in the protocol of the so-called "suspect" drugs and of which a very high percentage of patients are wary makes the initial approach immediately favorable. Lastly, the patient will have the correct perception of being followed throughout the entire course of treatment, a factor that can be decisive, especially in the presence of a pathology that has major psychological repercussions.

This multimodal approach provides the physician with a therapeutic protocol that can treat AGA in a broad, comprehensive, and at the same time highly targeted manner. This practice has been shown to rest on solid scientific foundations. Nevertheless, many aspects of the mechanisms underlying androgenetic alopecia (AGA) remain unclear. It is believed, however, that this clinical activity may help bridge some of these gaps by providing stem cell research with much-needed practical support. In conclusion, we can therefore say that the combination of ART and photobiostimulation represents an innovative and effective protocol in the treatment of androgenetic alopecia, with consistent and noteworthy results (Figs. 6.9, 6.10, 6.11, 6.12, 6.13, 6.14, 6.15, 6.16, 6.17, 6.18, 6.19, 6.20, 6.21, 6.22, 6.23, and 6.24).

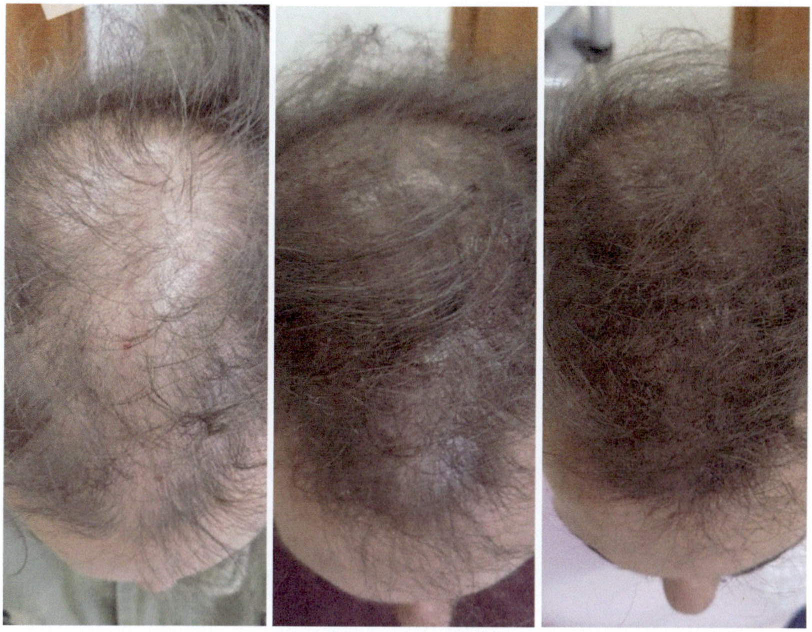

Fig. 6.9 Hair regrowth progression shown in three top-view images, illustrating transition from significant thinning to fuller scalp coverage over time

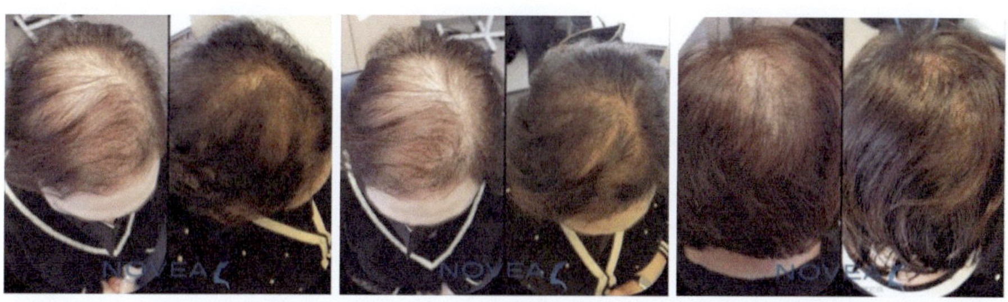

Fig. 6.10 Four sequential top-view images demonstrating progressive improvement in hair density, from initial significant thinning to fuller coverage over time

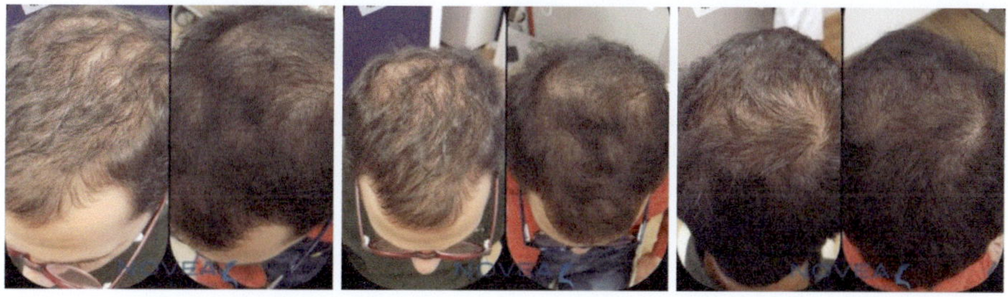

Fig. 6.11 Sequential top-view images before and after hair restoration treatment, illustrating progressive increase in hair density over time

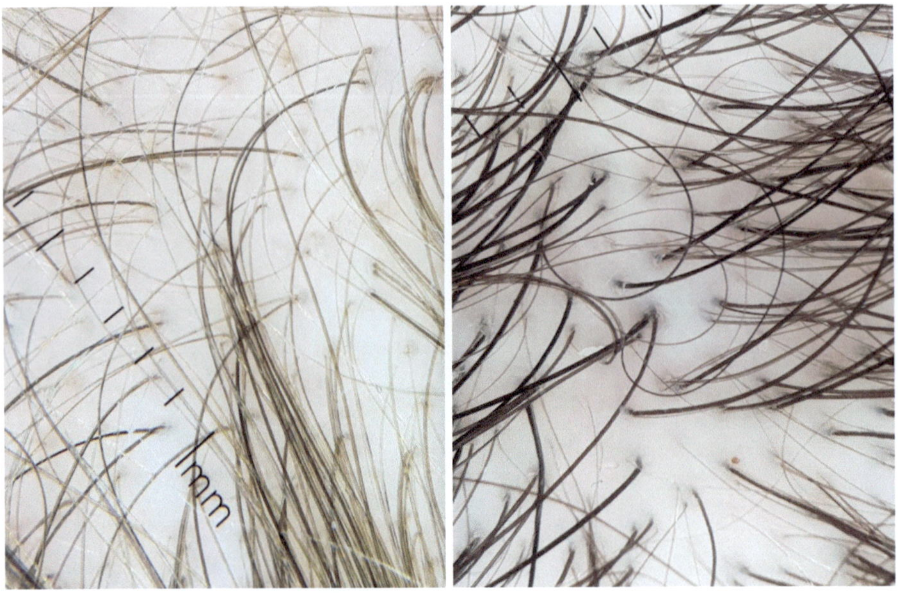

Fig. 6.12 Microscopic comparison of hair types: lighter, finer strands on the left and darker, thicker strands on the right. A "1 mm" scale provides size reference, highlighting structural differences

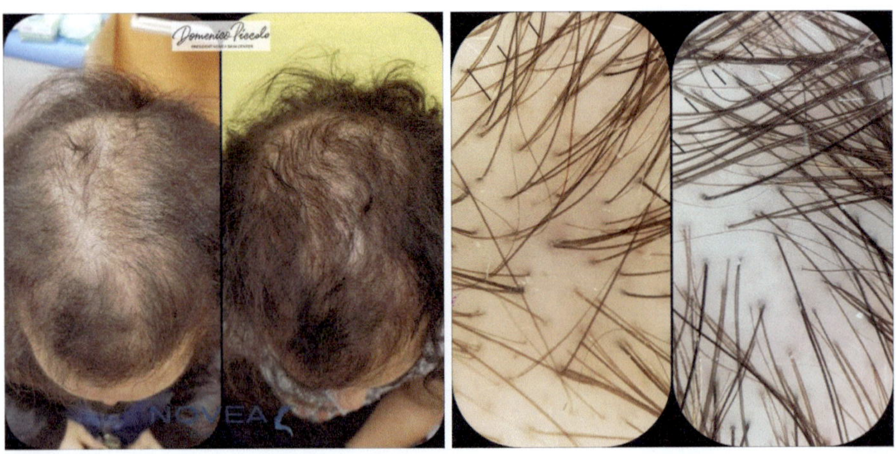

Fig. 6.13 Side-by-side comparisons illustrating hair restoration progress. Left: top-view scalp images showing visible regrowth. Right: close-up of hair follicles demonstrating increased density

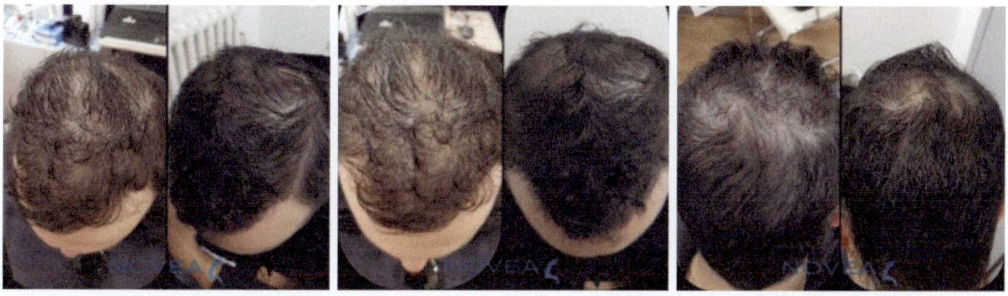

Fig. 6.14 Sequential inclined view—front view—rear view images showing progressive hair regrowth, with visible improvement in scalp coverage and hair density over time

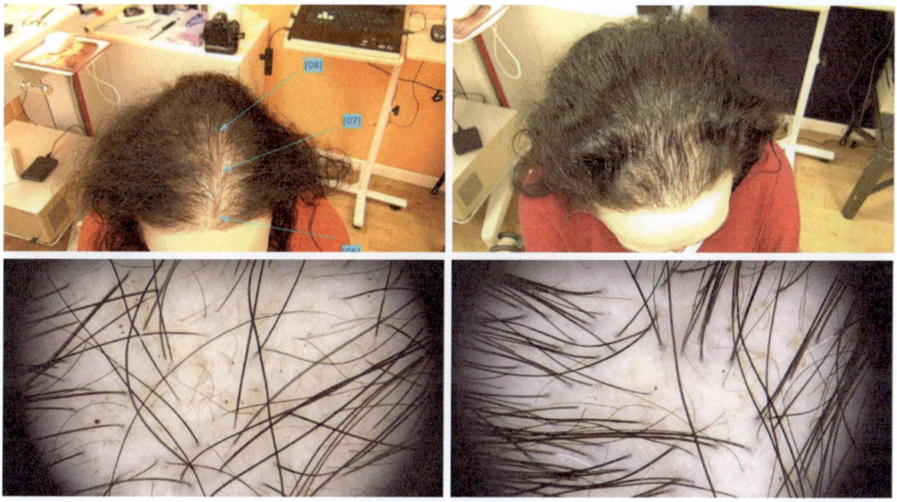

Fig. 6.15 Comparative top-view and close-up images of two women's scalps highlighting hair thinning. Blue labels mark points of interest on the top left image. The clinical setting underscores the diagnostic context of the examination

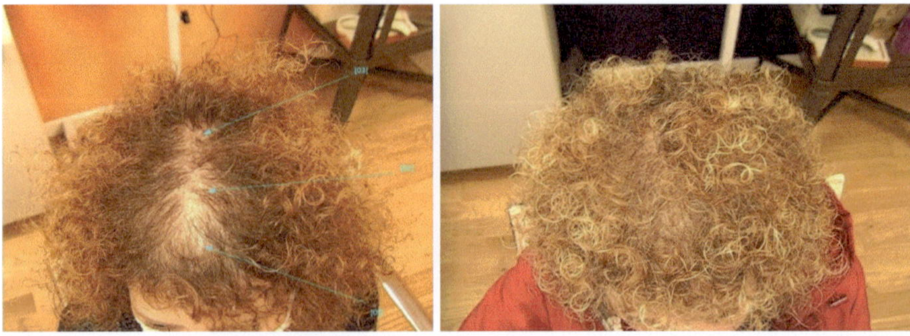

Fig. 6.16 Top-view comparison of curly hair before and after treatment. The left image highlights thinning areas, while the right shows visibly fuller curls, indicating improved hair density

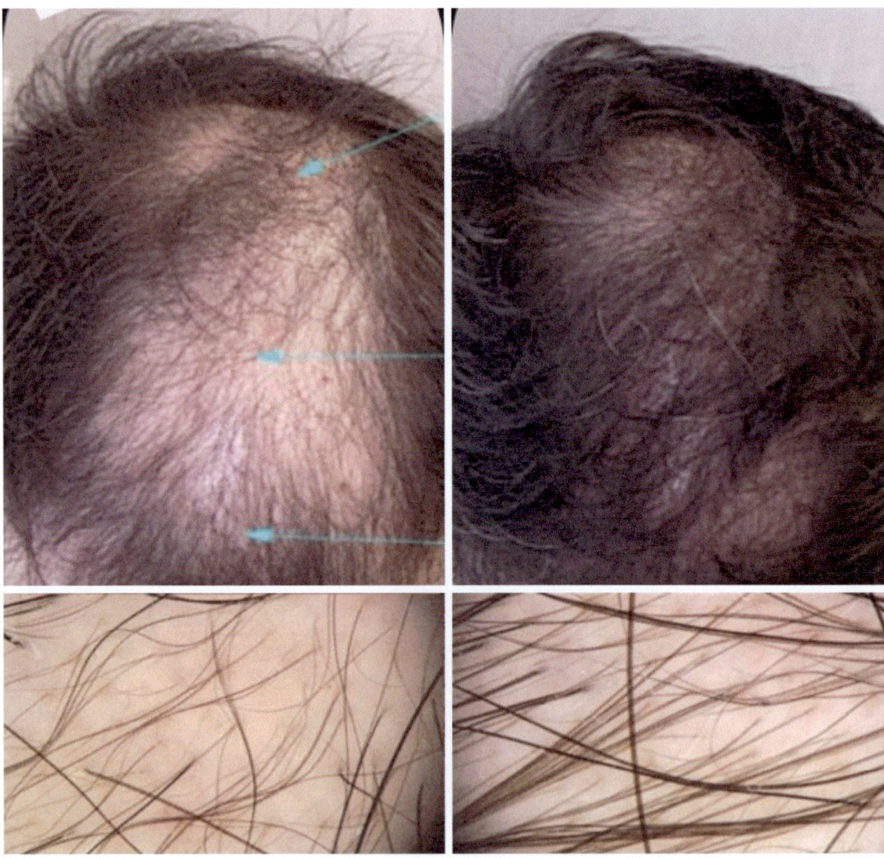

Fig. 6.17 Hair growth comparison showing scalp images before and after treatment (top row) and close-up views of individual strands (bottom row), illustrating increased hair density and thickness

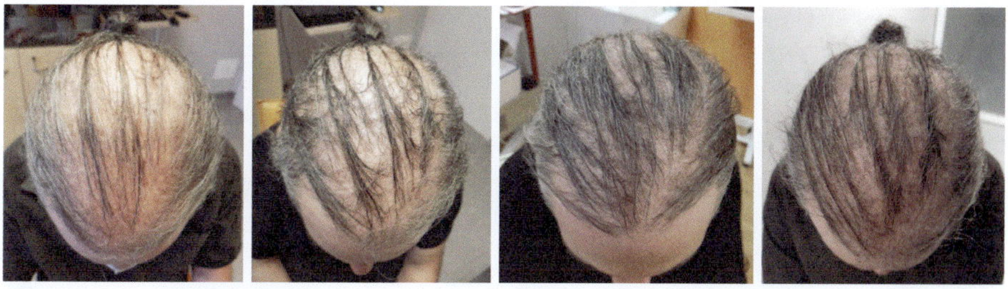

Fig. 6.18 Four top-view images documenting varying stages of hair thinning, highlighting progressive scalp visibility and suggesting a focus on hair loss evaluation or treatment outcomes

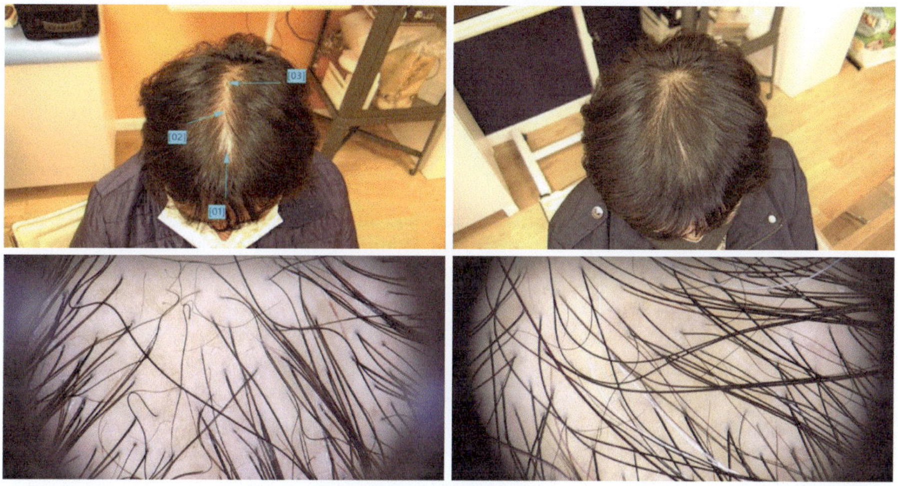

Fig. 6.19 Split-view image showing hair treatment results. Top: comparison of scalp before (with visible thinning and numbered markers 001–003) and after treatment (improved density). Bottom: close-up images highlighting hair growth progress

Fig. 6.20 Side-by-side comparison of scalp before and after treatment, showing transition from significant hair thinning to noticeable regrowth and increased hair density

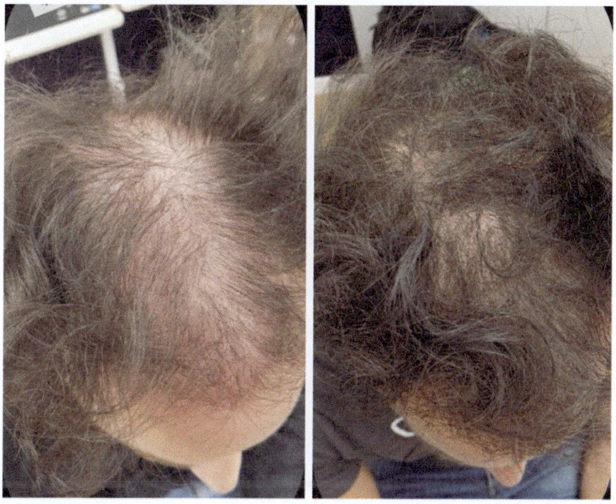

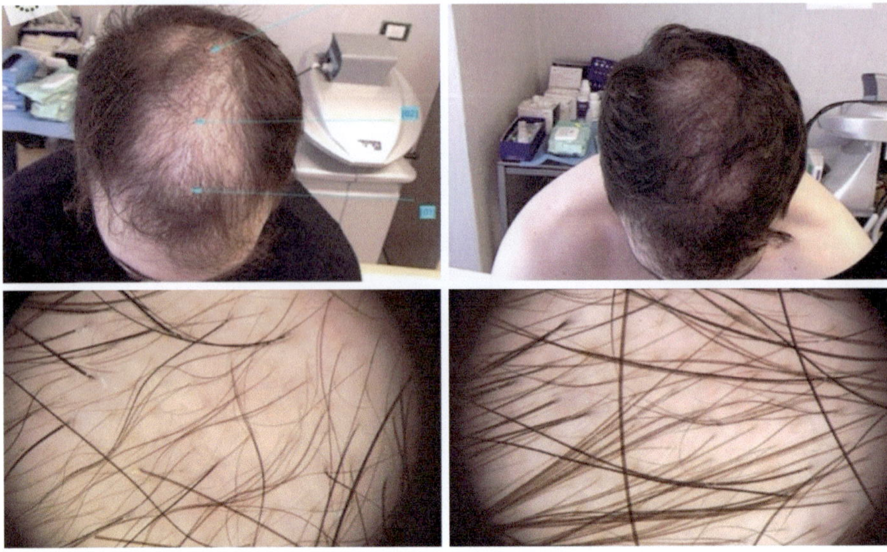

Fig. 6.21 Scalp and hair density comparison before and after treatment in a male patient. Top row: top-view images showing thinning versus regrowth. Bottom row: close-ups highlighting increased density. Medical equipment in the background indicates a clinical setting

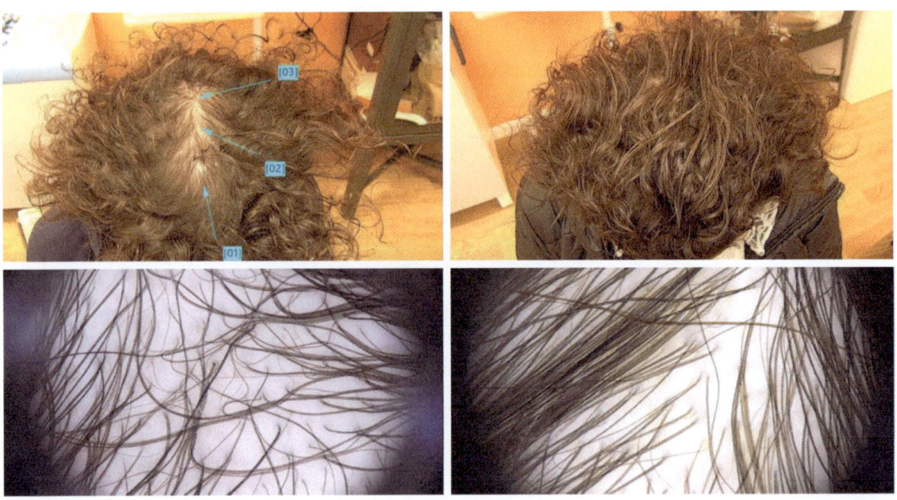

Fig. 6.22 Split image illustrating hair analysis. Top: scalp with curly hair marked by labels (001–003) and corresponding top view showing curl density. Bottom: microscopic close-ups highlighting hair texture and strand density in a clinical or salon setting

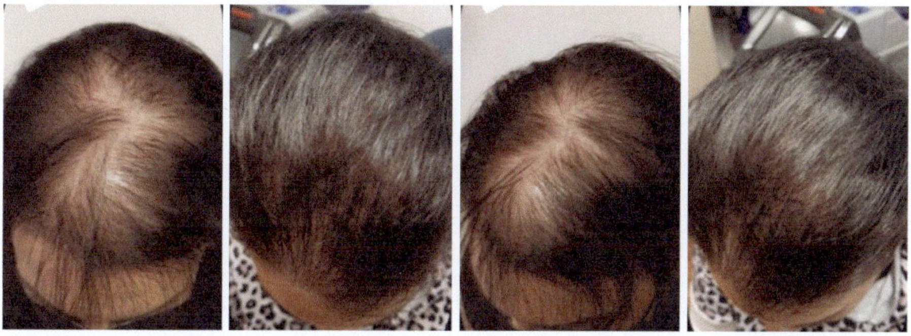

Fig. 6.23 Four sequential images showing progressive hair growth. Alternating images highlight improvement in hair thickness over time, with medical equipment in the background suggesting a clinical treatment setting

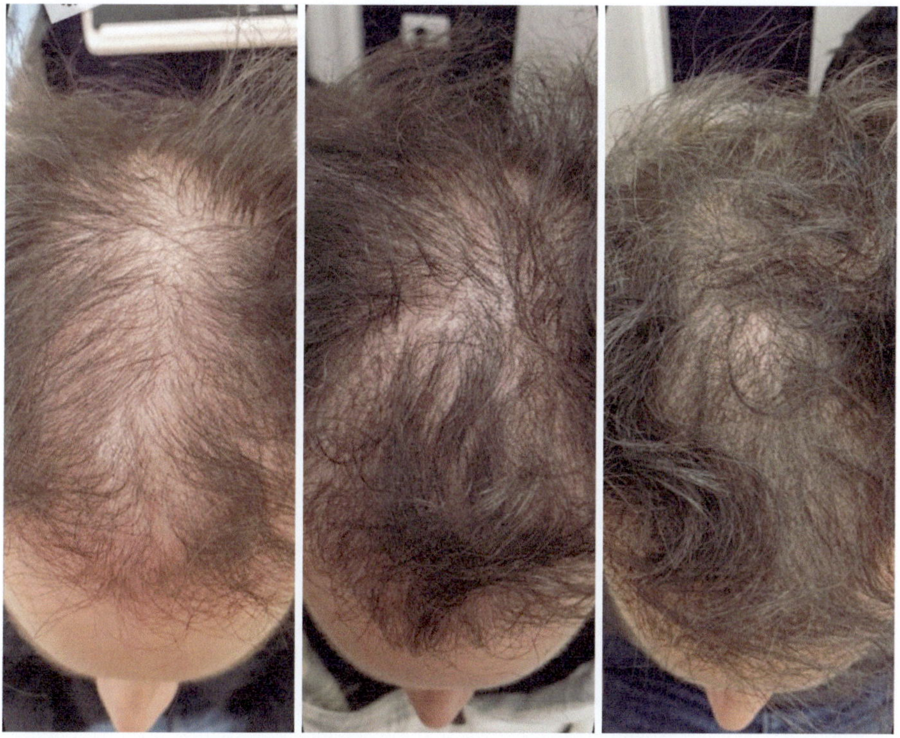

Fig. 6.24 Three sequential top-view images demonstrating progressive hair regrowth, from significant thinning to moderate regrowth and ultimately fuller scalp coverage

References

1. Kabir Y, Goh C. Androgenetic alopecia: update on epidemiology, path ophysiology, and treatment. J Egypt Womens Dermatol Soc. 2013;10(3):107–16.
2. Piraccini BM, Alessandrini A. Androgenetic alopecia. G Ital Dermatol Venereol. 2014;149(1):15–24.
3. Kaliyadan F, Nambiar A, Vijayaraghavan S. Androgenetic alopecia: an update. Indian J Dermatol Venereol Leprol. 2013;79(5):613–25.
4. Inui S, Nakajima T, Itami S. Significance of dermoscopy in acute dif fuse and total alopecia of the female scalp: review of twenty cases. Dermatology. 2008;217(4):333–6.
5. Seleit IA, El-Bakry OA, El-Sherif RM. Dermoscopy in hair disorders. Menoufia Med J. 2014;27:762–5.
6. Ludwig E. Classification of the types of androgenetic alopecia (common baldness) occurring in the female sex. Br J Dermatol. 1977;97(3):247–54.
7. Moftah N, Moftah N, Abd-Elaziz G, et al. Mesotherapy using dutasteride-containing preparation in treatment of female pattern hair loss: photographic, morphometric and ultrustructural evaluation. J Eur Acad Dermatol Venereol. 2013;27:686–93.
8. Sobhy N, Aly H, El Shafee A, El Deeb M. Evaluation of the effect of injection of dutasteride as mesotherapeutic tool in treatment of androgenetic alopecia in males. Our Dermatol Online. 2013;4:40–5.
9. Barbareschi M. The use of minoxidil in the treatment of male and female androgenetic alopecia: a story of more than 30 years. G Ital Dermatol Venereol. 2018;153(1):102–6.
10. Gajjar PC, Mehta HH, Barvaliya M, Sonagra B. Comparative study between mesotherapy and topical 5% minoxidil by dermoscopic 8of8 evaluation for androgenic alopecia in male: a randomized controlled trial. Int J Trichol. 2019;11(2):58–67.
11. Abdallah MA, El-Zawahry KA, Besar H. Mesotherapy using dutasteride-containing solution in male pattern hair loss: a controlled pilot study. J Pan Arab Leag Dermatol. 2009;20:137–45.
12. Khorosandi K, Hosseinzadeh R, Abrahamse H, Fekrazad R. Biological responses of stem cells to photobiomodulation therapy. Curr Stem Cell Res Ther. 2020;15(5):400–13.
13. Tosti A, Whiting D, Iorizzo M, et al. The role of scalp dermoscopy in the diagnosis of alopecia areata incognita. J Am Acad Dermatol. 2008;59(1):64–7.
14. Pillai JK, Misore V. Role of low-level light therapy (LLT) in androgenetic alopecia. J Cutan Aesthet Surg. 2021;14(issue 4)
15. Lubart R, Eichler M, Lavi R, et al. Low-energy laser irradiation pro motes cellular redox activity. Photomed Laser Surg. 2005;23(1):3–9.
16. Smith AN, Willis E, Chan VT, et al. Mesenchymal stem cells induce dermal fibroblast responses to injury. Exp Cell Res. 2010;316:48–54.
17. Rossi MAF, Ricci F, Vignoli F, et al. In vitro multilineage potential and immunomodulatory properties of adipose derived stromal/stem cells obtained from nanofat lipoaspirates. CellR4 4 (2016): e2212.
18. Rossi M, Roda B, Zia S, et al. Characterization of the tissue and stromal cell components of micro-superficial enhanced fluid fat injection (micro-SEFFI) for facial aging treatment. Aesthet Surg J. 2020;40:679–90.
19. Ogura F, Wakao S, Kuroda Y, et al. Human adipose tissue possesses a unique population of pluripotent stem cells with nontumorigenic and low telomerase activities: potential implications in regenerative medicine. Stem Cells Dev. 2014;23:717–28.
20. Tesauro P, Trivisonno A, Gennai A, et al. Hair transplantation in cicatricial alopecia: the role of autologous fat transfer. Int J Regen Med. 2020;3:11.
21. Bernardini FP, Gennai A. Superficial enhanced fluid fat injection for volume restoration and skin regeneration of the periocular aesthetic unit. An improved fat grafting technique to enhance the beauty of the eye. JAMA Plastic Facial Surgery. 2016;18:1.

22. Gennai A, Tesauro P, Colli M, et al. Infranatant portion of microfragmented adipose tissue: a promising source of SVF for the management of androgenic alopecia. Int J Regen Med. 2021;12:1–7.
23. Gennai A, Bovani B, Colli M, et al. Comparison of harvesting and processing technique for adipose tissue graft: evaluation of cell viability. Int J Regen Med. 2021;4:2–5.
24. Gennai A, Bernardini FP, Baldessin M, et al. Evaluation of the viability and Phenotipe of adipose derived cells harvested using different harvesting and processing procedures: a pilot study. J Stem Cells Clin Pract. 2022;2:101.
25. Amore AL, Bartoletti E, Gennai A. The role of regenerative medicine in androgenetic alopecia treatment: an uncontrolled clinical trial. Aesthetic Med. 2022;8:457.
26. El-Khalawany M, Rageh MA, Elnokrashy I, et al. Efficacy of autologous stroll vascular fraction injection in the treatment of androgenic alopecia. Arch Dermatol Res.
27. Piccolo D, Crisman G, Conforti C, et al. Trichobiolight: a new, effective protocol in the treatment of androgenetic alopecia and telone effluvium. Dermatol Ther. 2021;34:e14799.
28. Piccolo, D Tesauro, P Conforti, C Crisman, G Tramontozzi, S Bovani, B Melfa, F Gennai, A (2023) (2023) A New Office Based Protocol to Reverse AGA Miniaturization at Early and Advanced Stages Journal of Surgery and Research 06(03) 10.26502/jsr.001010.26502/jsr.10020313.

Chapter 7
Cell-Based Strategies in Reconstructive Hair Transplantation: Regeneration and Remodeling of the Scalp After Injury

A key focus is placed on hair transplantation feasibility in scarred tissue, addressing critical factors such as recipient site preparation, donor area evaluation, follicular unit survival, and graft retention. The chapter discusses technical nuances of performing follicular unit extraction (FUE) and strip harvesting (FUSS) on scarred areas, highlighting the limitations and advantages of each approach.

Clinical case studies showcase the effectiveness of combining Autologous Regenerative Therapy with hair transplantation, demonstrating improved graft survival rates, enhanced tissue integration, and superior aesthetic outcomes. The findings underscore the importance of stem cell-based regenerative strategies in maximizing the success of hair restoration in complex cases of scarring alopecia.

The chapter concludes by discussing long-term graft stability, risk minimization, and the future of regenerative approaches in reconstructive trichology, positioning guided SEFFI Stem & Vol as a pioneering solution in hair transplantation for scarred scalp lesions.

7.1 Introduction

The evolving perception of the human body and its evident social repercussions deeply affect individuals who perceive themselves as lacking aesthetic appeal. This often results in insufficient personal and relational well-being, sometimes escalating into genuine psychological distress.

In this context, hair plays a central role, and hair loss represents a psychological challenge far more significant than an aesthetic one. Those affected often embark on an intense search for solutions. Among these, hair transplantation (HT) has emerged as one of the most sought-after therapeutic options and is a rapidly expanding specialization.

D. Piccolo et al., *A.R.T. Autologous Regenerative Therapy in Aesthetic Medicine*, https://doi.org/10.1007/978-3-032-05835-5_7

When hair loss is caused by trauma, surgical procedures, or disease leading to the development of scarred, hairless areas, the sudden or rapidly progressive nature of these conditions often results in distress surpassing that caused by common baldness. This is frequently due to the overtly pathological appearance of such scars.

Even in these cases, hair transplantation has rightfully become one of the most effective solutions for addressing cicatricial alopecia. Global statistics reveal a marked increase in HT procedures, with 2% being performed on scarred areas (source: ISHRS, International Society of Hair Restoration Surgery). However, patients with cicatricial alopecia often face difficulties in identifying available reconstructive options.

Even when they locate highly skilled specialists, they are likely to encounter varying treatment plans. This reflects the diverse training and approaches among practitioners in this field. The author believes a multidisciplinary approach is essential for treating these patients effectively and with minimal trauma. In addition to medical specialists, patients often turn to professionals in hairpiece design or scalp micro-pigmentation (SMP), which offers a quick and increasingly common camouflage solution, particularly for men.

The abundance of options can lead to significant uncertainty and fear of making the wrong choice. In extreme cases, this may result in a paralyzing confusion, and it is not uncommon to encounter patients who have lived with their scars for years. For patients to make informed and confident decisions, they need a thorough understanding of the reconstructive treatment options and their limitations.

This chapter highlights a reconstructive approach that combines autologous adipose tissue transplantation with hair transplantation, proposing it as an innovative and effective solution.

For many years, the relative merits of autologous scar transplantation compared to skin expansion have been debated, as the reliability of these two procedures appeared to differ. Moreover, only a small number of surgeons were skilled in these techniques, and the results often fell short of expectations regarding the number and quality of transplanted hair. Today, hair transplantation is the leading choice for addressing most cases of cicatricial alopecia. Previous limitations have been expanded and must be redefined.

In recent years, advancements in surgical techniques, combined with regenerative medicine procedures, have helped overcome justified fears of failure. These developments have led to far more predictable and consistent outcomes for hair transplantation in this context.

Several factors have historically hindered progress in this field. Chief among them are the belief that scar tissue is unsuitable for hosting grafts and the absence of such options within public healthcare systems.

Today, we know that performing hair transplantation on scarred areas requires adapting to a distinct environment. Understanding, measuring, and minimizing these differences significantly reduces the unpredictability of autotransplantation outcomes.

This chapter provides a detailed exploration of the complexities of this approach.

7.2 Content Overview

7.2.1 Cicatricial Scalp Lesions: Primary and Secondary Forms

This section delves into the complexity of cicatricial scalp lesions, which are categorized into primary and secondary forms. These classifications result in heterogeneous scar "terrains" with significant histological differences between the two types. For instance, primary lesions are often characterized by the absence of hair and mild scalp thickness atrophy. In contrast, secondary lesions—caused by the underlying traumatic or pathological event—may lead to scars with tissue deficiency and a scalp of reduced thickness, varying significantly from case to case.

This section introduces the principles of hair transplantation, focusing on the calculations required to determine the intervention's minimum feasibility. Historically, surgeons worked on compromised recipient areas with poor vascularization and lower tissue quality compared to normal tissue. However, they relied on the follicular unit's inherent ability to sufficiently improve scar tissue.

7.2.2 Recipient Area Preparation with the Guided SEFFI Method

In the past, hair transplantation performed directly on scalp scars often disappointed both patients and surgeons. Over time, transplanted follicles in scarred areas tended to thin, leading to unsatisfactory outcomes. This section details how preparing the recipient area with the guided SEFFI technique can address this issue.

7.2.3 Clinical Applications and Surgical Experiences

This extensive section bridges theoretical concepts with clinical practice, presenting the results of years of surgical experience. The numerous cases demonstrate that improving compromised tissue quality creates ideal conditions for subsequent hair transplantation, maximizing results. By presenting clinical experience, this chapter aims to define the reasons for and methods of using this regenerative technique across various forms of cicatricial alopecia.

7.2.4 Scarring Alopecia

7.2.4.1 Definition and Classification of Scarring Alopecias, Including Primary and Secondary Forms

The term *scarring alopecia* refers to a clinical condition characterized by the permanent loss of hair follicles and, consequently, hair, regardless of the causative factor. The most significant feature of scarring alopecia is the irreversibility of the pathological process due to the complete destruction of the follicular bulge, the area where stem cells are located. This is often accompanied by tissue atrophy and sclerosis, resulting in definitive alopecia.

The prior existence of a pilosebaceous unit is a fundamental requirement for defining an alopecia as scarring, along with the presence of a destructive inflammatory process and the formation of a permanent scar [1, 2].

Generic Causes

- Genetic and developmental defects.
- Inflammatory/autoimmune diseases.
- Infectious conditions.
- Neoplasms.
- Metabolic-degenerative disorders.
- Idiopathic conditions.

Primary Scarring Alopecias

Primary scarring alopecias are defined as those conditions in which the fibrotic and scarring outcome is the result of a heterogeneous group of pathological processes that ultimately destroy the pilosebaceous unit.

When there is sufficient clinical, histopathological, and experimental evidence demonstrating the presence of an inflammatory insult specifically targeting the hair follicle, the condition is referred to as primary or follicular scarring alopecia.

- Below are the most common forms of primary or follicular scarring alopecias:
- Lichen planopilaris.
- Discoid lupus erythematosus.
- Folliculitis decalvans.
- Brocq's pseudopelade.

All other forms are rare, rarely diagnosed, or nosologically uncertain. In some cases, differential diagnosis between the various forms—both clinically and histopathologically—can be challenging, if not impossible. This difficulty reflects the limited understanding of the etiological factors and pathogenic mechanisms underlying these diseases.

In the pathogenesis of acquired primary scarring alopecies, the topographic site of perifollicular inflammation plays a significant role. When critical areas, such as the infundibulum and the follicular bulge, are involved, scarring alopecia results.

The classification of primary scarring alopecies is complex and multifaceted. However, based on the histological nature of the perifollicular cellular infiltrate, acquired primary forms can be broadly categorized into lymphocytic and neutrophilic types [1, 2].

- Acquired Primary Lymphocytic Scarring Alopecias:
- Lichen planopilaris.
- Graham-Little syndrome.
- Frontal fibrosing alopecia.
- Scarring alopecia associated with androgenetic alopecia.
- Discoid lupus erythematosus.
- Brocq's pseudopelade.
- Alopecia parvimaculata.
- Central centrifugal cicatricial alopecia.
- Follicular degeneration syndrome.
- Acquired Primary Neutrophilic Scarring Alopecias:
- Folliculitis decalvans.
- Tufted hair folliculitis.
- Acne keloidalis nuchae.
- Erosive pustular dermatosis of the scalp.
- Dissecting cellulitis.

These conditions result in more or less diffuse scarring of the scalp.

Pharmacological treatment options aimed at halting the progression of the disease and mitigating the associated inflammatory process include intralesional, systemic, and topical medications.

Scarring alopecias can therefore be highly heterogeneous. Nevertheless, mesenchymal stem/stromal cells (MSCs) have been employed for years in this domain for their regenerative capacity and potential therapeutic benefits [4].

Notably, MSCs are now known to induce remodeling of scar tissue while modulating inflammatory processes, reducing the risk of disease reactivation. In simple terms, the goal is to minimize the risk of relapse, enabling these lesions to be treated similarly to secondary scarring alopecias.

A correct diagnosis is crucial for promptly initiating therapeutic measures to arrest the progression of the pathological process and prevent further expansion of hairless, scarred areas. Accurate photographic and videodermoscopic documentation will facilitate critical assessment of the clinical picture and determine whether the condition has stabilized.

Once the pathological process has been halted, typically after a 2-year stabilization period, surgical intervention may be considered to restore hair presence in the existing scarred areas.

Autologous Regenerative Therapy can play a pivotal role in mitigating the aforementioned risks. Specifically, the use of the stromal vascular fraction (SVF), rich in adipose-derived mesenchymal stem cells (AD-MSCs), has shown exceptional promise. These cells possess significant potential to differentiate into various cell types while exerting anti-inflammatory, antiapoptotic, and pro-angiogenic effects [5]. Moreover, their secretome—a vast array of bioactive factors they release—positively influences wound healing, promoting new blood vessel formation, cell proliferation, and inflammation modulation [5].

In particular, treatments involving nanofat grafting, which deliver high concentrations of AD-MSCs and exosomes, may prove especially beneficial for managing the inflammatory mechanisms characterizing primary scarring alopecias. This therapeutic approach has the potential to significantly improve the predictability and long-term stability of hair transplant outcomes while reducing the risk of recurrence. Clinical studies have demonstrated promising results, especially in conditions like lichen planopilaris [3].

Although hair transplantation can provide aesthetically pleasing and natural results, it is not without risks. The possibility of disease reactivation following the procedure poses a challenge, necessitating a cautious approach and careful monitoring.

7.2.4.2 Secondary Scarring Alopecias

Secondary scarring alopecias, or non-follicular scarring alopecies, occur when the destructive process does not specifically target the pilosebaceous unit but involves it nonspecifically.

This category includes alopecies resulting from trauma that causes skin injury with consequent tissue damage and permanent loss of skin appendages. Traumatic insults may be mechanical, thermal, radiant, chemical, bacterial, or viral. The damaged hair structures are replaced by fibrous tissue as a result of the scarring process [1, 2].

The scarring process restores the anatomical continuity of the affected area through tightly interconnected stages, including inflammation, granulation tissue formation, and subsequent remodeling of the connective matrix. Imbalances in collagen production and enzymatic breakdown during these stages can lead to the development of hypotrophic, hypertrophic, or, less commonly, keloid scars.

In secondary scarring alopecies, unlike primary forms, there is no risk of autoimmune or inflammatory reactivation. However, the affected tissue is often more severely compromised and depleted. In these cases, regenerative medicine techniques can be particularly beneficial in enhancing skin thickness and trophism, preparing the recipient site for hair transplantation.

The absence of an underlying autoimmune pathological process in secondary scarring alopecies is advantageous, as it limits the risk of recurrence after treatment. Nonetheless, the structural and functional impairment of the affected skin presents specific challenges that must be addressed to achieve satisfactory and long-lasting results.

7.3 Paragraph 2: Indications for Hair Transplantation in Patients with Scarring Alopecia

Determining a patient's suitability for a hair transplantation procedure on a scarred scalp lesion involves considering three main factors that are essential prerequisites for achieving a satisfactory aesthetic result. It is crucial to present the patient with a high likelihood of a positive outcome, especially given that they may already have endured the trauma of sudden or distressing hair loss. Without a favorable prognosis, the patient may be inclined to explore less invasive alternatives, such as prosthetics, thereby avoiding a potentially burdensome and challenging reconstructive process.

The three primary factors to consider are as follows:

- The extent of the scarred area.
- The density of the donor area.
- The diameter and characteristics of the hair.

In addition to these elements, other aspects should be evaluated, such as the quality of the scar tissue, the potential for effective regenerative treatments, and the patient's genuine willingness to undergo the procedure. Additional considerations include the evolutionary risk associated with scarring alopecia, particularly if linked to a classic form of androgenetic alopecia (AGA), and the availability of alternative surgical options, such as direct excision, tissue expansion, or prosthetic solutions.

Each case has unique characteristics that require thorough and personalized analysis. In this context, adopting an approach based on rigorous data and measurements is essential. For many years, and not just in scarring forms, the guiding principle in this field has remained: "measure everything."

7.3.1 Extent of the Recipient Area

The size of the alopecic scarred area should always be carefully measured and compared with the donor area's capacity.

Whatever the availability of follicles for harvesting, this is always the starting point for hypothesizing various reconstructive scenarios.

The ideal perspective for any patient is complete coverage of the bald area. When this is not possible, partial coverage options can be considered to achieve significant aesthetic impact. For example, reconstructing the hairline or the anterior area, combined with strategically planned haircuts and styling, can deliver very satisfactory results.

Residual scarred areas can be compensated for, particularly in women, with prosthetic implants.

Determining whether "the effort is worth the reward" requires study and experience, but the following cases aim to demonstrate how the balance of choice is rapidly shifting from major procedures, such as scalp reductions and expanders, to this minimally invasive technique.

Hair transplantation offers significant technical advantages: it allows precise control of follicle direction, ensures immediate recovery of work and social activities, and has fewer complications [9] and lower costs.

7.3.2 Density of the Donor Area

The fundamental requirement for undergoing surgical hair thickening is having a good donor area density (occipital, parietal, and temporal zones). If donor area density is low, appropriate assessments of the patient's age and predisposition to baldness (androgenetic alopecia) must be made before proceeding with surgery. This involves determining whether it is worthwhile to use an already limited donor area and, if so, the timing and method for doing so.

Ectopic donor areas, such as the beard in men, can also be considered.

7.3.3 Hair Diameter and Characteristics

Hair diameter and characteristics significantly influence transplant results and, consequently, the degree of coverage achieved. Hair thickness, color, and texture can substantially affect outcomes, even with the same number of implanted follicular units (FUs).

For instance, wavy or curly hair provides greater coverage than straight hair. Additionally, lighter-colored hair tends to contrast less with the scalp's lighter tone, optimizing aesthetic appearance. Similarly, thicker hair (with a diameter exceeding 80 micrometers) ensures superior coverage compared to thinner hair.

7.3.4 Coverage Value

The *Coverage Value* (CV) is a method used to measure the total donor capacity (TDC) of a patient's temporal, parietal, and occipital areas during the preliminary evaluation. The TDC determines the maximum number of FUs that can be harvested over one or more procedures while avoiding depletion of the donor area.

The CV evaluation methodology follows a defined sequence:

- Analysis of the average hair caliber.
- Measurement of follicular unit density per square centimeter.
- Assessment of the average number of hairs per FU across different sectors.

These data, entered into specialized software, generate the TDC, which is then compared to the extent of the recipient surface.

The basic formula for calculating the CV is the product of FU density, the average number of hairs per FU, and the average hair caliber.

$$\text{CV} = \frac{\text{FU}}{\text{cm}^2} \times \frac{\text{hairs}}{\text{FU}} \times \text{caliber}$$

The Coverage Value (CV) calculation was developed by Dr. Koray Erdogan as an objective method to standardize scientific criteria in hair analysis. For calculating Coverage Value (CV) in hair analysis: CV = (follicular units/cm^2) × (hairs per FU) × (hair calibre), representing a composite measure of hair density and thickness.

A CV of 5.4 represents the minimum coverage value required to achieve a satisfactory aesthetic result. Recently, two apps have been introduced to increase the speed of data processing.

These tools allow the determination of the number of grafts required to obtain a satisfactory outcome and help decide whether it is advisable to fully or partially cover the scarred area. In cases where partial coverage is preferred, the available resources are used to transplant only the areas of primary aesthetic importance.

7.4 Paragraph 3: Guided SEFFI Stem & Vol Pretransplantation Technique

7.4.1 Premise

Regenerative treatments utilizing adipose clusters and stem cells derived from adipose tissue aim to stimulate regenerative processes and restore lost volume in tissues. Their application is widespread in reconstructive surgery across various specialties, with specific techniques tailored to adapt these treatments to the type of tissue in which they are employed.

In this chapter, we highlighted how hair transplantation techniques have improved through the adoption of precise procedures, ensuring the integrity and vitality of follicular units (FUs) from extraction to placement in the recipient site.

FUs are composite grafts, thin but metabolically active, requiring robust nutrient diffusion from the recipient site after implantation. When the structure of the recipient site is damaged or altered—whether due to primary or secondary causes—it is essential to enhance its quality before proceeding with transplantation.

To address this on a scarred scalp, characterized by unique anatomical and physiological traits, a tailored and adaptable Autologous Regenerative Therapy (ART) technique has been developed. This approach accommodates the varying characteristics of the previously described scar lesions.

7.4.2 The S&V Stem & Vol Technique

The name of the guided SEFFI Stem & Vol (S&V) technique reflects the sequential use of the stem cell component followed by the adipose component. The name was inspired by the tennis action "Serve & Volley," which, when executed correctly, can be decisive with just two strokes.

The S&V technique involves three fundamental steps:

- Harvesting.
- Processing.
- Grafting.

7.4.2.1 The Harvesting Process

The adipose tissue harvesting procedure is identical to the guided SEFFI technique [6], previously described in other chapters. It is a guided technique performed under local anesthesia. The fat is collected using micro-perforated, micro-hooked cannulas with a diameter of 2 mm and perforations of 0.8 mm, significantly smaller than traditional cannulas (Fig. 7.1).

The volume of anesthetic infiltrated and adipose tissue aspirated varies depending on the size of the scarred area to be treated. In all cases treated by the author, the volume has ranged between a minimum of 10 cc and a maximum of 30 cc.

7.4.2.2 The Processing Phase

This phase distinguishes the S&V technique from the traditional guided SEFFI method. Only part of the harvested adipose tissue is subjected to emulsification and centrifugation (Fig. 7.2), while another portion is kept intact [6].

7.4.3 Application of the Two Fractions

The use of the two fractions varies depending on the type of scar tissue, making it impossible to establish a rigid protocol. The more atrophic and thinner the scar, the greater the need to expand it and increase its thickness. The fluidity of the vascular stromal fraction is ideal for effectively performing this mechanical action while simultaneously providing a robust biological effect.

This phase facilitates the subsequent step: the infiltration of the volumetric component composed of plurifragmented adipose tissue. Numerous studies demonstrate that the stem cell component enhances the performance of adipose tissue, reducing

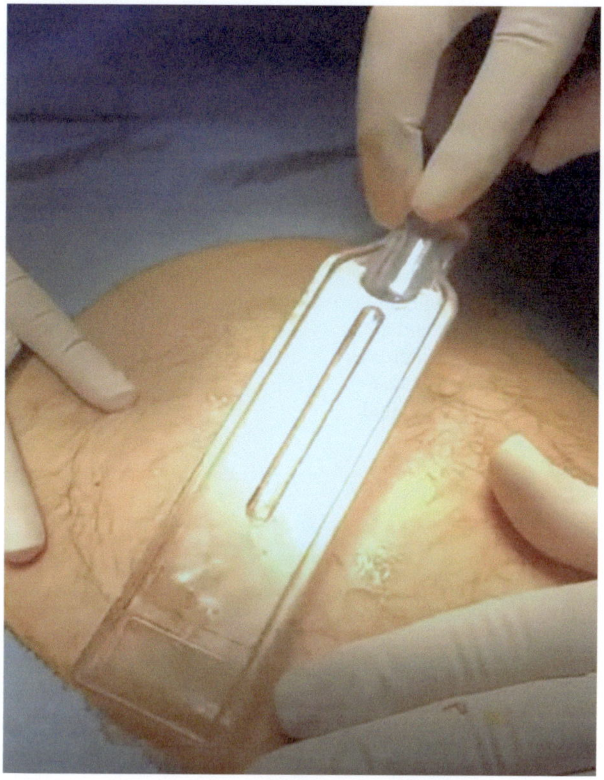

Fig. 7.1 Autologous Regenerative Therapy using the guided SEFFI device for precise instrument alignment. A gloved professional operates within a sterilized field, with a blue drape surrounding the treatment site, highlighting its application in minimally invasive procedures

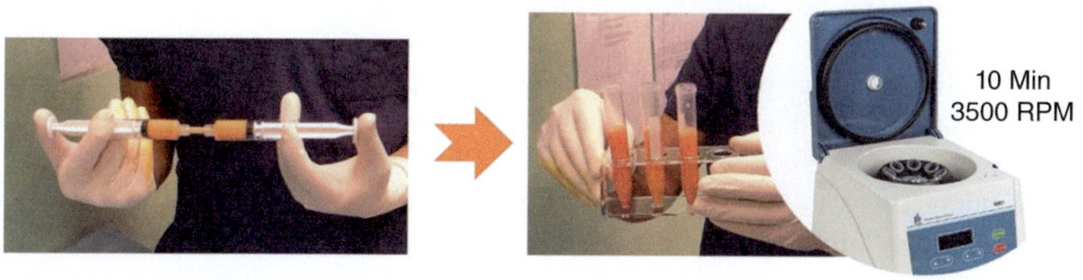

Fig. 7.2 Stepwise preparation of a stromal vascular fraction graft using the guided SEFFI protocol. Left: a syringe containing the graft is held preprocessing. Middle: contents are transferred into tubes for centrifugation. Right: centrifuge set at 3500 RPM for 10 min to separate the infranatant portion from adipocytes

its rate of resorption. However, this rate remains significant in cases of severe scar atrophy, making it necessary to consider a single session often insufficient.

On average, the volume of vascular stromal fraction injected per square centimeter of scar tissue ranges from **0.3 to 0.5 ml**, while the volumetric component volume injected per square centimeter ranges from **0.2 to 0.4 ml** (Fig. 7.3).

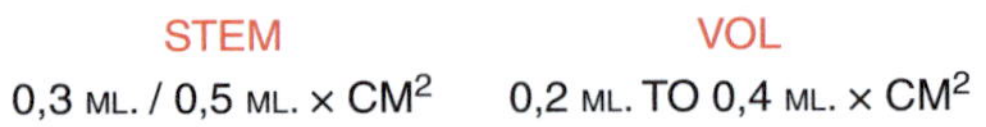

Fig. 7.3 Comparison of volume measurements per square centimeter for two treatment categories. "STEM" indicates 0.3–0.5 mL/cm^2, while "VOL" represents 0.2–0.4 mL/cm^2, highlighting dosing guidelines for different procedural approaches

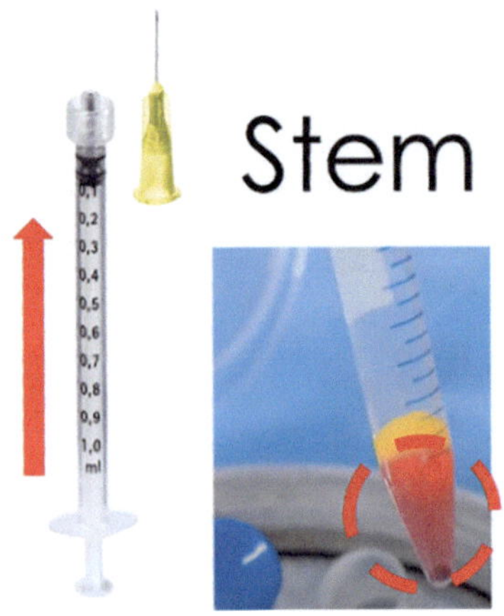

Fig. 7.4 Visual representation of a "Stem" preparation step. A syringe marked 0–1 mL with an upward red arrow and a yellow needle cap indicates sample aspiration. A test tube containing red liquid, highlighted with red semicircles, suggests the targeted portion for extraction

7.4.3.1 The Grafting Phase

The cellular fraction of the graft, rich in vascular stromal and stem cells (STEM), is transferred into 1 ml Luer lock syringes and injected using 30G needles to expand the scar tissue. This approach promotes a multilayer distribution of the cellular fraction (Fig. 7.4).

The volumetric fraction (VOL) is similarly transferred into 1 ml Luer lock syringes but is injected into the pre-expanded tissue using 21G or 23G needles in the subcutaneous layer.

It is important to note that scar tissue has a profoundly altered anatomy, meaning the precision of the "infiltration plane" is often theoretical, particularly in cases of severe atrophy. In practice, the adipose tissue is injected into the thin available layer of tissue.

7.4.3.2 Injection Sites and Technique

Injection sites are carefully planned to ensure complete coverage of the scarred area while minimizing their number (Fig. 7.5). This approach reduces the risk of adipose tissue leaking from prior entry points during infiltration, especially in compact scar tissue.

Using needles as the infiltration tool allows for precise, minimally invasive, and atraumatic injection of microfragments across multiple layers. To mitigate thromboembolic risks, the fat is released in very small quantities using a retrograde technique.

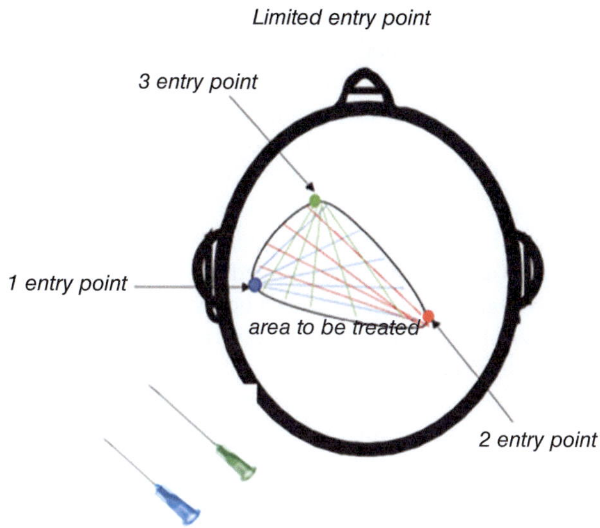

Fig. 7.5 Diagram illustrating the "Limited entry point" technique. Three colored entry points (blue, green, red) connect to a central treatment area, with corresponding colored needles shown below, representing strategic access routes for targeted interventions

The fragility of many scar tissues makes the use of rigid microcannulas hazardous, as they could tear the tissue. Conversely, the compact nature of scar tissue often prevents the release of adipose tissue with flexible microcannulas, further validating the use of needles for this procedure.

This method, currently under publication, reduces the risks associated with classical lipostructure and allows for the modulation of injected quantities, tailoring them to the specific needs of the tissue. It is a variation of the "recharging with ADSc" techniques for infiltrated fat and appears to be a simple and effective approach for treating scar lesions on the scalp.

The S&V technique significantly improves the survival rate of adipose tissue grafts, restoring essential functions to the scalp necessary for performing a hair transplant with a high margin of safety.

7.4.3.3 Number of Procedures

Procedures using the S&V technique are typically spaced at intervals of approximately 3 months. In our clinical experience, the number of sessions is almost always limited to two, with rare exceptions requiring three sessions.

The positive effects of hair transplantation in cases of scarring alopecia have been known for years. The *Strip* technique was predominantly used for these cases until 2015. It offered the possibility of maintaining usual hairstyles, and since it was preferred by the majority of surgeons even for traditional transplants on healthy scalp tissue, it appeared to be the ideal technique. Grafts derived from a Strip retained a greater amount of perifollicular adipose tissue, offering the possibility of implanting a well-structured composite graft.

This apparent advantage was at the same time the limitation of this approach. Follicular units richer in adipose tissue required more nourishment to ensure their survival, which in scar tissue necessitated performing conservative, low-density transplants. In such circumstances, the surgeon, relying solely on experience and common sense, had no tools to determine a safe density for the placement of FUs.

After 2015, the advent of a modern and refined FUE technique brought with it the advantage of more versatile use of the donor area. Follicular units derived from FUE, which on average were thinner and more skeletonized than those obtained through the Strip method, immediately seemed to perform better. This minimally invasive technique also favored better postoperative recovery by removing less tissue overall from the scalp and offering the possibility of using ectopic donor areas such as the beard.

Despite this evolution, hair transplantation on scarred lesions often produced variable results. Numerous successes were counterbalanced by unexplained medium-term failures. This led to a degree of caution in proposing hair transplantation as a valid alternative to traditional reconstructive plastic surgery techniques, limiting its use to cases where scars covered only small percentages of the scalp [7, 8].

Failures or partial successes resulted in very different scenarios, such as a reduced percentage of hair growth, thinner caliber of the grown hair, or a follicular cycle unable to renew properly.

This led to the idea of enriching scar tissue beforehand to maximize the chances of success of subsequent transplants, thus avoiding the loss or damage of some cellular components crucial for follicular regeneration. This gave rise to the guided SEFFI Stem & Vol technique.

Beyond any theoretical conviction regarding the advantages of this procedure, one of the greatest practical demonstrations of its effectiveness lies in the comparison between the results of cases treated "tout court" with hair transplantation alone and others treated with the combined S&V + hair transplantation approach. Although there are no scars or clinical situations that are perfectly identical, several cases with similar characteristics have been examined, and all comparisons have tangibly highlighted the advantages of this new methodology.

7.5 Match 1: Details

The cases compared involve long-standing burn injuries. Both showed a partial reduction in scalp thickness. Both patients had an excellent donor area.

In the case from October 2009, the woman underwent a hair transplant using the Strip technique, with approximately 1600 FUs positioned. The 2023 case was treated with the FUE technique, with 1450 FUs positioned.

The outcome was positive in both cases but with a different graft survival rate, estimated at around 70% in the first case and 100% in the second case. In this comparison, certain technical advancements previously described undoubtedly played a role in favor of the second case, but the overall superior result is evident (Fig. 7.6).

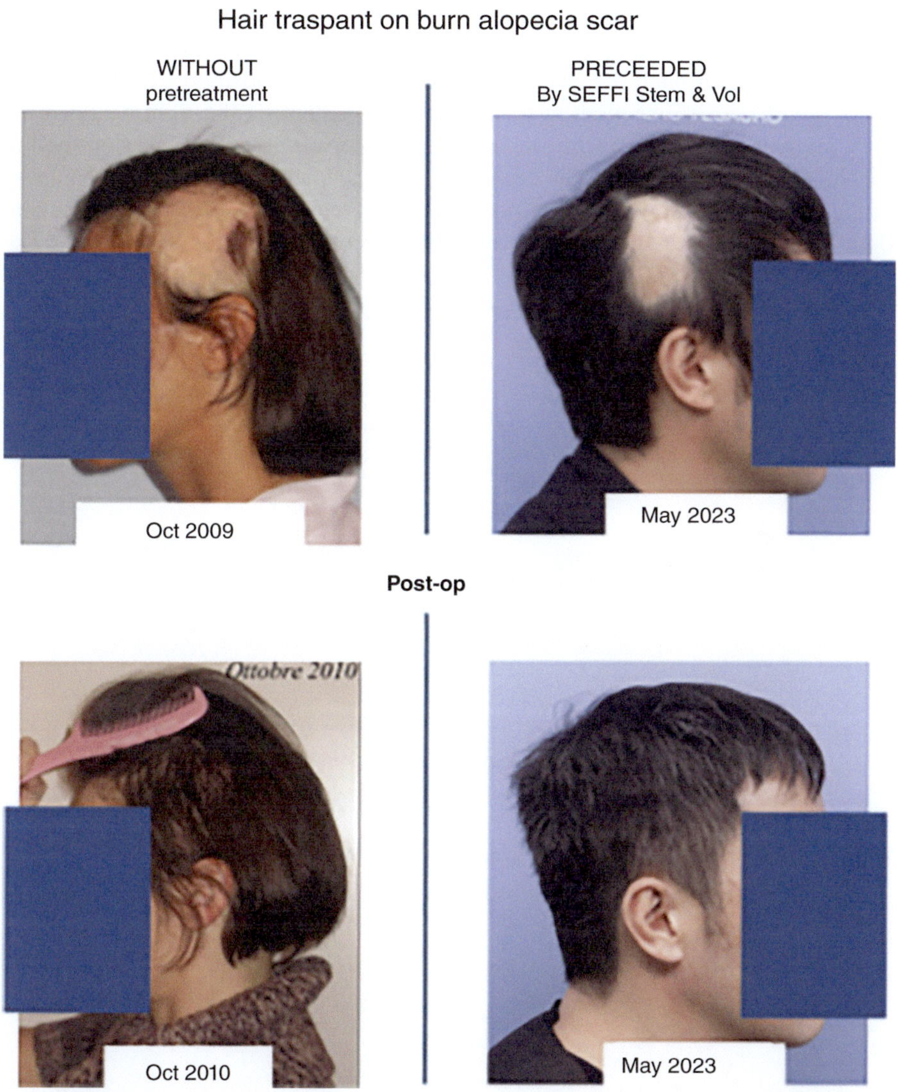

Fig. 7.6 Comparison of hair transplant outcomes on burn alopecia scars. Top left and bottom left: a female patient without pretreatment shows moderate improvement posttransplant (2009–2010). Top right and bottom right: a male patient pretreated with SEFFI Stem & Vol demonstrates enhanced hair coverage posttransplant (2023), suggesting improved outcomes with regenerative pretreatment

7.6 Match 2: Details

The cases compared involve long-standing burn injuries, highlighting the significant difficulty these patients have faced for years in finding a viable reconstructive solution. Unlike the previous cases, these were treated within a short time of each other (2022 and 2023), both using the FUE technique with a similar graft density in the recipient areas, totaling 2219 and 2507 follicular units (FUs), respectively.

Both cases exhibited a reduction in scalp thickness and marked fibrosis. The donor area hair was excellent in both cases. The patient on the right declined the S&V pretreatment, whereas the second patient underwent two S&V sessions before the hair transplant. Although the density of FU placement in the recipient area was similar in both cases, there was a noticeable difference in graft survival rates and slightly poorer hair quality in the patient who did not undergo the S&V (Fig. 7.7).

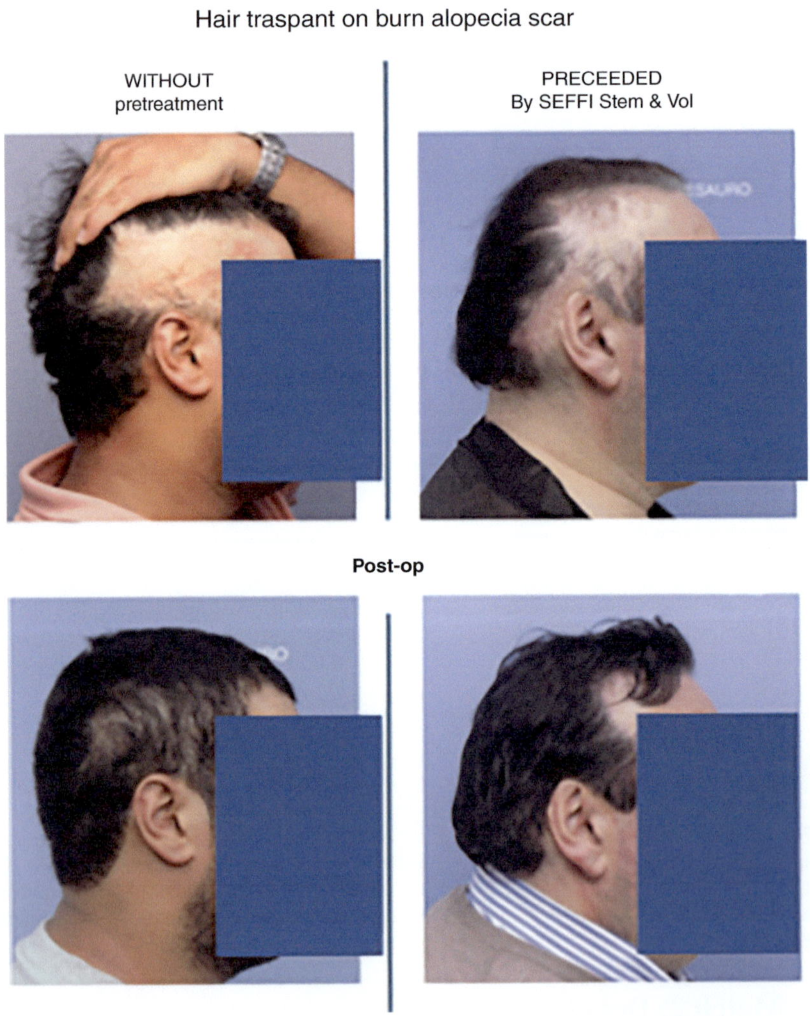

Fig. 7.7 Comparative results of hair transplantation on burn alopecia scars. Top row: pretreatment images showing cases without (left) and with (right) SEFFI Stem & Vol pretreatment. Bottom row: postoperative outcomes reveal improved hair coverage in both, with enhanced results in the pretreated case. Faces are obscured by logos for anonymity

The variability of results in interventions without S&V pretreatment has been significantly reduced, showing excellent stability of outcomes over time.

This evolution has transformed our approach to patients with scalp scar lesions. Recognizing the significant benefits of proper pretreatment of scar tissue, we have expanded the indications for this approach. As will be explored in the next section, these techniques have rightly begun to replace many more invasive and costly reconstructive plastic surgery procedures.

Moreover, we have observed that performing two or more pretreatment sessions facilitates hair transplantation, enabling a graft density in the recipient area comparable to that of healthy tissues and reducing the need for multiple sessions.

The main advantages achieved can be summarized in three key points:

- Better reproducibility and consistency of results.
- Reduced procedural risks.
- Improved postoperative recovery.

7.6.1 Technical Notes on Hair Transplantation on Scar Tissue

Hair transplants on previously grafted scar lesions must consider four factors that significantly differentiate them from transplants on healthy tissue:

- Intense ischemic response to vasoconstrictors.
- Reduced graft retention by the tissue.
- Increased fragility of the skin.
- Greater vulnerability to infections.

Below, these aspects are examined in detail:

Under normal conditions, adipose tissue can withstand prolonged ischemia, but recent experimental studies show that under stress, it releases inflammatory molecules. In cases of subsequent reperfusion, these molecules can cause arteriolar vasoconstriction, leading to stasis and blockage of capillary flow. These secondary phenomena can compromise FU viability. Hence, it is essential to avoid or significantly limit the use of vasoconstrictors in the recipient area, restricting their use to peripheral application.

AFT procedures can restore and regenerate the thickness and consistency of scar tissue, reducing the need to excessively tilt the follicles during placement. However, a drawback is the reduced graft retention capacity of the tissue due to a decrease in the number of elastic fibers. This can result in "popping" during placement. The best prevention involves conducting multiple preliminary tests to determine the exact proportion between incisions and available grafts. To this end, it is important to have a wide range of blades or implantors of various sizes. Additionally, using the "stick and place" technique is preferable for placing "oversized" grafts (single or wide three-unit grafts).

Although the skin improves in quality with adipose tissue treatments, it loses stability due to a weaker connection with deeper structures. This resulting abnormal mobility over the underlying planes can make it more prone to tearing during graft placement. The technique must be adapted accordingly. To enhance the visibility of the surgical field, multiple sprays of saline solution are preferable to gauze padding.

The increased vulnerability to infections, due to the absence of skin appendages, necessitates adopting sterility measures for this procedure that are not usually required in standard hair transplantation. All patients are required to start a preventive antibiotic regimen. A brief video summary of the described method is available online.

7.7 Paragraph 4: Guided SEFFI Stem & Vol and Autotransplantation: Clinical Cases

Over the past 6 years, I have applied the Stem & Vol technique in combination with autotransplantation in a total of 53 patients affected by scarring alopecia of the scalp.

Among these, five patients presented with primary cicatricial alopecia. Specifically, two patients suffered from primary lesions resulting from systemic autoimmune diseases, such as Graft-versus-Host Disease (GVHD) and Romberg syndrome/scleroderma—rare conditions that impair tissue regeneration and destroy hair follicles.

The remaining 46 patients were affected by cicatricial lesions of secondary origin.

To date, 48 patients have completed the full therapeutic cycle and are currently under posttreatment observation. The detailed analysis of these cases has provided valuable insights into individual therapeutic responses, allowing further refinement of the treatment protocol.

All comorbidities were carefully assessed and monitored using dermatoscopy and, when necessary, appropriately treated prior to initiating surgical phases. In particular, for patients with primary cicatricial alopecia, a 2-year waiting period was observed from the point the condition was considered stable and clinically "inactive."

This process ensured optimal patient preparation, creating the ideal conditions for successful treatment outcomes.

S&V procedures significantly contributed to the improvement in both quality and quantity of residual hair in healthy peripheral areas surrounding the cicatricial lesions. The results were markedly positive, with improvements observed in the aesthetic appearance, functionality, and resilience of the transplanted hair—crucial elements for patients' psychological wellbeing.

Below, we present a series of detailed clinical cases, including patient information, diagnoses, treatments carried out, clinical observations, and photographs supporting the outcomes achieved. This material will provide an in-depth overview of the efficacy of the adopted technique and the improvements obtained in the management of cicatricial alopecia.

7.8 Scarring Alopecia: Case 1: Lichen Planopilaris Scar

Patient: Male, 64 years old.
Diagnosis: Lichen planopilaris.
Treatments:

- 2020: S&V session
- 2020: FUE with 1310 FUs
- 2023: FUE with 1200 FUs.

Case Description:

The patient, a 60-year-old man, was diagnosed with predominantly oral lichen planopilaris, which later manifested on the scalp. An intense effluvium preceded the formation of scar tissue in the frontoparietal region.

In 2020, an S&V treatment was performed to prepare the scalp for a subsequent transplant aimed at increasing the density in the anterior and central areas. This procedure involved the placement of 1310 follicular units and was scheduled after a 3-month interval.

Only 4 months after the FUE transplant, the situation stabilized and remained so for the following 3 years without significant progression of the lichen planopilaris.

To further improve density, a second FUE transplant was planned and carried out in 2023, with 1200 follicular units harvested and transplanted (Fig. 7.8).

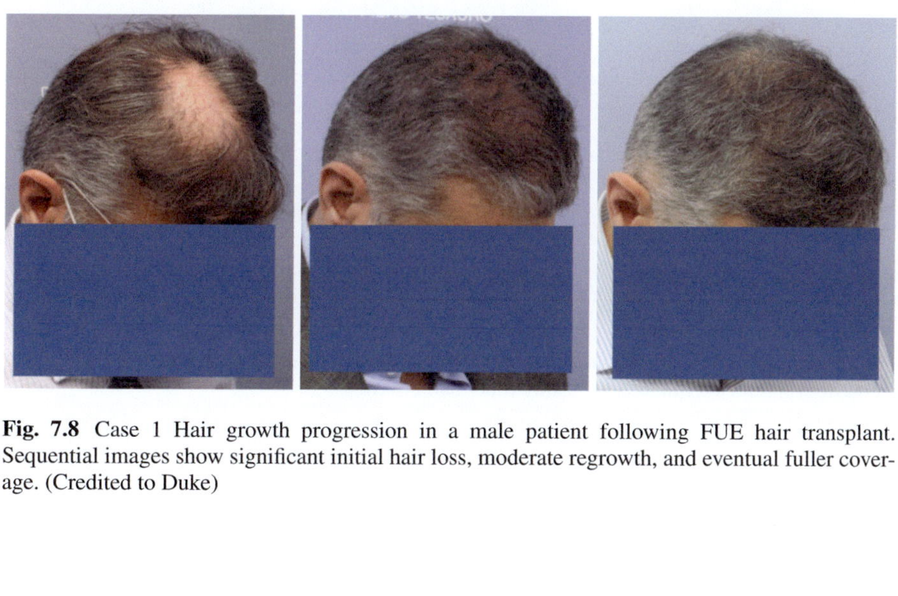

Fig. 7.8 Case 1 Hair growth progression in a male patient following FUE hair transplant. Sequential images show significant initial hair loss, moderate regrowth, and eventual fuller coverage. (Credited to Duke)

7.9 Scarring Alopecia: Case 2: Scarring Alopecia Due to Scalp Infection

Patient: Female, 35 years old.
Diagnosis: Probable pseudotinea amiantacea with infectious complications in early childhood.
Treatments:

- Four skin expansion surgeries.
- January 2022: first S&V session.
- May 2022: second S&V session.
- 2022: FUE with 1611 FUs.

Case Description:

The patient's parents observed an intensive and progressive loss of hair starting at the age of 3. The diagnosis was a fungal infection, never confirmed, which did not respond to treatment and led to the development of scar tissue in the frontoparietal area.

At the age of 18, the patient underwent her first skin expansion surgery, followed by three more, all yielding poor results. At the age of 31, she underwent a regenerative treatment with S&V, followed by a hair transplantation procedure.

The results, achieved through FUE with 1611 follicular units, were excellent and are evident in the photographs below (Fig. 7.9).

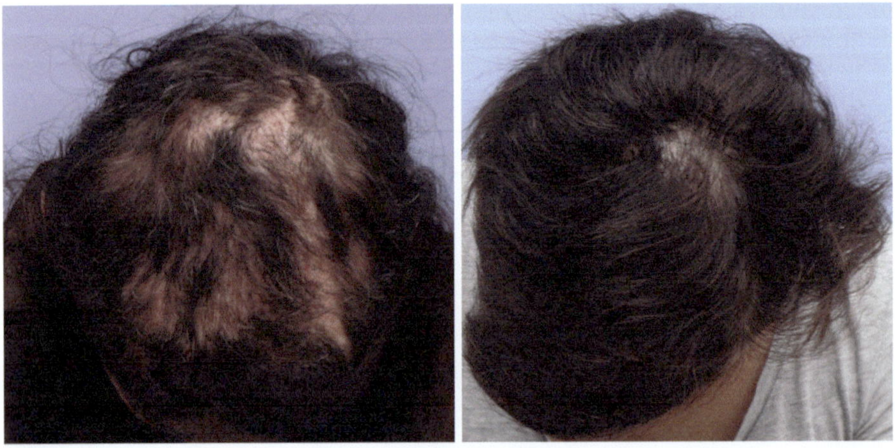

Fig. 7.9 Case 2 Side-by-side comparison of scalp before and after hair treatment, showing a transition from significant thinning to visibly fuller hair coverage

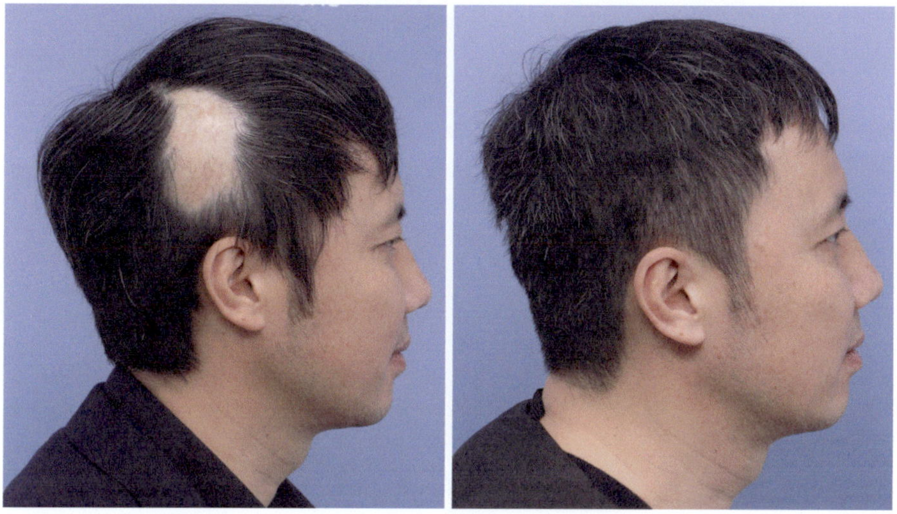

Fig. 7.10 Case 3 Profile view comparison before and after hair transplant. Left: significant scalp hair loss. Right: restored hair density posttreatment. (Credited to GOT IT Transplant)

7.10 Scarring Alopecia: Case 3: Burn-Related Scarring Alopecia

Patient: Male, 44 years old.
Diagnosis: Scarring alopecia caused by a burn from boiling water at the age of 1.
Treatments:

- July 2023: S&V session.
- October 2023: FUE with 1450 FUs.

Case Description:

The images below demonstrate the results achieved with a combined treatment of AFT (autologous fat transfer) and hair transplantation on a scar in the temporoparietal area (Fig. 7.10).

The scar resulted from a burn caused by boiling water in early childhood.

7.11 Scarring Alopecia: Case 4: Burn-Related Scarring Alopecia

Patient: Male, 46 years old.
Diagnosis: Burn injury at the age of 4 (boiling oil)—70% of the scalp burned.
Treatments:

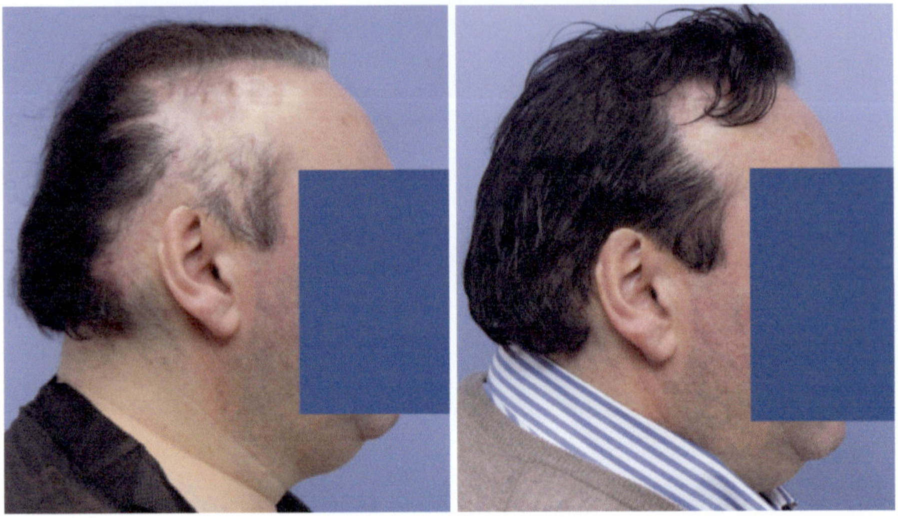

Fig. 7.11 Case 4 Side profile comparison of a male patient before and after hair restoration, showing improvement from significant hair loss to fuller hair coverage

- May 2022: first S&V session.
- September 2022: second S&V session.
- March 2023: FUE with 2507 FUs.

Case Description:

The images below illustrate another case of scarring alopecia caused by a burn to the scalp due to contact with boiling oil at the age of 4 (Fig. 7.11).

Eighteen years after the incident, the patient underwent a skin expansion procedure, which resulted in a partial reduction of the scarred area.

The hair transplant, however, effectively restored the alopecic area.

The surgical thickening was preceded by two S&V sessions, both including volumetric and cellular components, performed 4 months apart.

7.12 Scarring Alopecia: Case 5: Burn-Related Scarring Alopecia

Patient: Male, 36 years old.
Diagnosis: Scarring from a burn caused by a pot of boiling oil at the age of 1.
Treatments:

- February 2019: first S&V session.
- March 2019: second S&V session.
- May 2019: FUE with 1427 FUs.

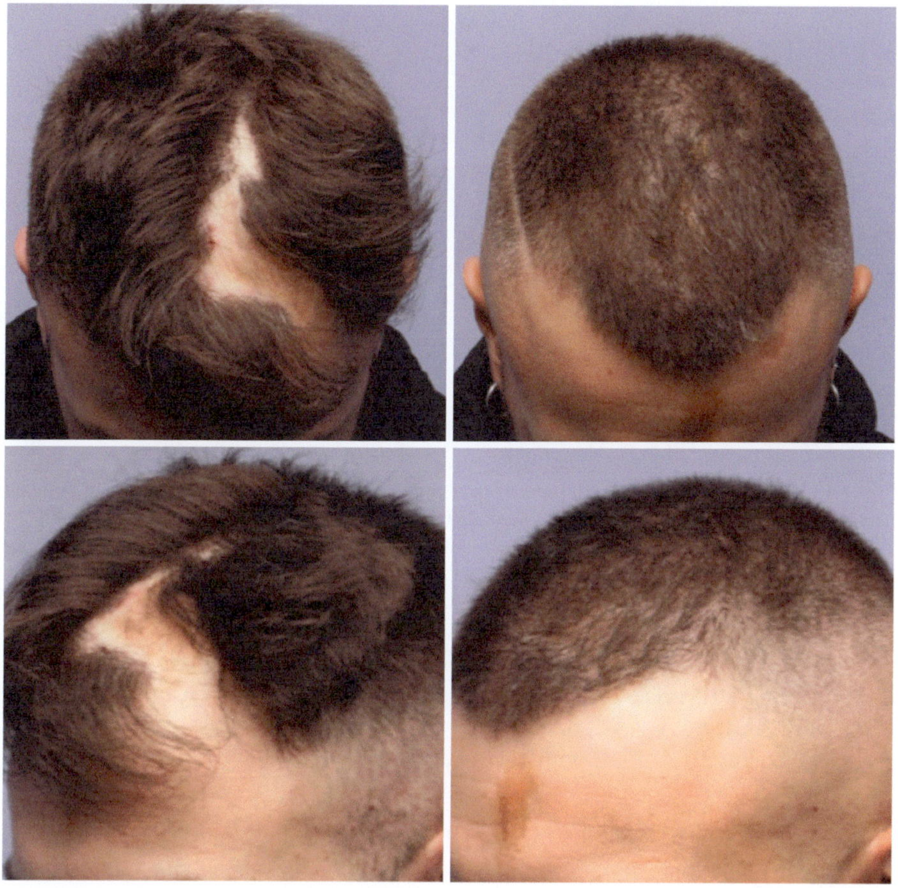

Fig. 7.12 Case 5 Top-view comparison before and after hair restoration procedure, illustrating a transition from thinning hair and visible scalp to noticeably fuller hair coverage. (Credited to Shapiro Medical Group)

Case Description:

This case demonstrates another instance of scarring alopecia resulting from a pediatric burn to the scalp.

The surgical thickening was preceded by two S&V sessions. The subsequent FUE hair transplant produced the results visible in the images below (Fig. 7.12).

7.13 Scarring Alopecia: Case 6: Scarring from Radiotherapy

Patient: Female, 32 years old.

Diagnosis: Scarring resulting from radiotherapy following surgery for the removal of Stage 4 multifocal Ewing sarcoma with subsequent metastases.

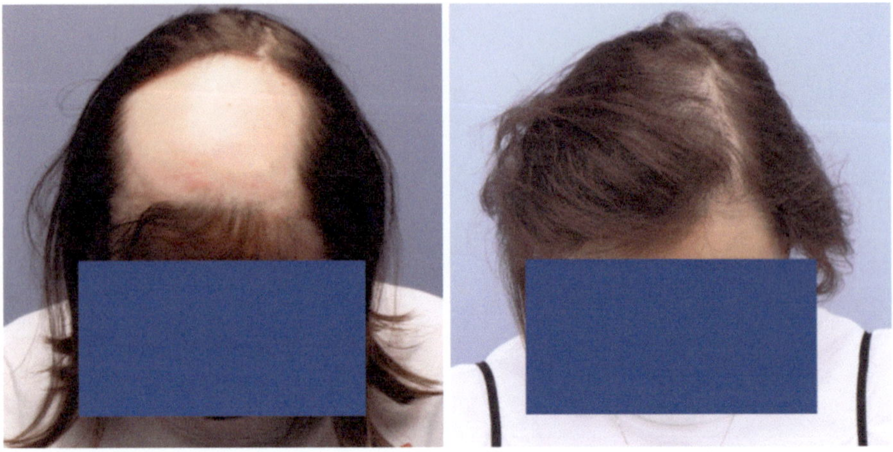

Fig. 7.13 Case 6 Hair growth comparison over time. Left image (June 2019) shows significant hair loss, while right image (July 2022) demonstrates fuller hair coverage following treatment

Treatments:

- February 2019: S&V session.
- June 2019: S&V session.
- October 2019: STRIP with 2900 FUs.
- May 2021: S&V session.
- November 2021: FUE with 1604 FUs.

Case Description:

The images below show the results achieved with a combined AFT (autologous fat transfer) and hair transplantation procedure on a scar in the frontal area (Fig. 7.13).

Before the FUE hair transplantation, a third regenerative S&V session was performed.

7.14 Scarring Alopecia: Case 7: Scarring from Radiotherapy

Patient: Female, 63 years old.

Diagnosis: Scarring resulting from surgical removal of a glioblastoma in 2007 followed by radiotherapy.

Treatments:

- May 2021: S&V session.
- November 2021: FUE with 1604 FUs.

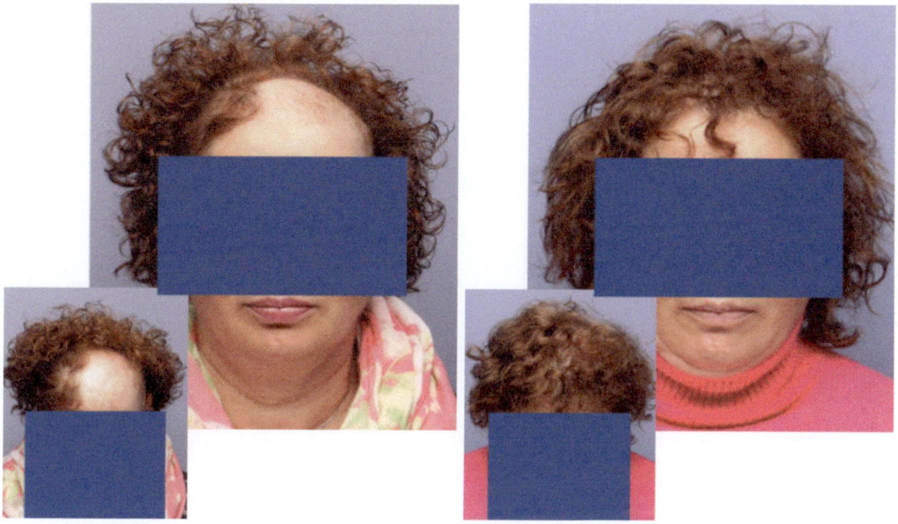

Fig. 7.14 Case 7 Before and after images showing hair restoration results. Top row highlights changes in the forehead and hairline, while bottom row presents overall facial appearance improvement following treatment

Case Description:

The patient, following glioblastoma removal surgery and subsequent scarring alopecia caused by radiotherapy, used a prosthetic implant for over 15 years. This further exacerbated the sclerosis of the scar tissue.

The regenerative treatment improved the tissue quality and the surrounding hair even before the hair transplant. However, the complete regrowth timeline for the transplanted hair was longer than expected (Fig. 7.14).

7.15 Scarring Alopecia: Case 8: Scarring from Radiotherapy

Patient: Male, 39 years old.

Diagnosis: Scarring resulting from radiotherapy following medulloblastoma treatment.

Treatments:

- February 2020: first S&V session.
- May 2020: second S&V session.
- July 2020: FUE with 1374 FUs.

Case Description:

The images below illustrate another case of scarring alopecia caused by radiotherapy, located in the occipital area.

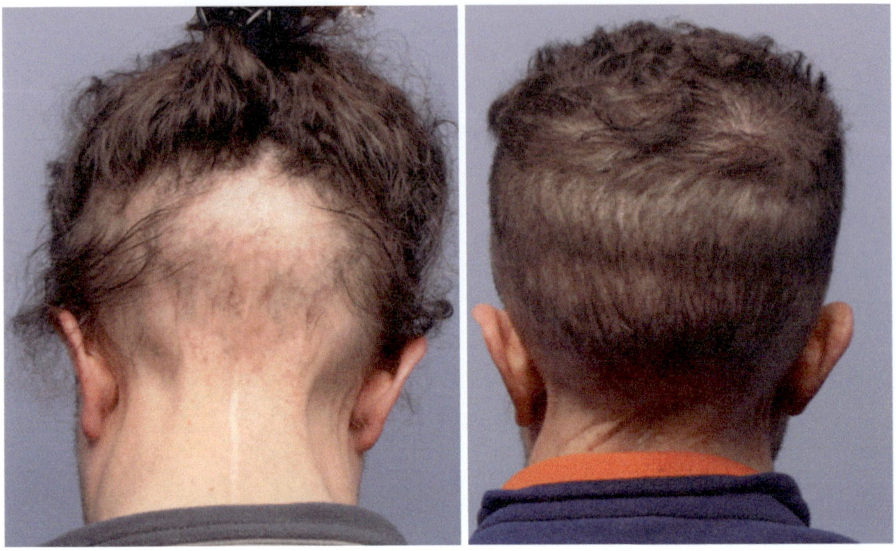

Fig. 7.15 Case 8 Back-of-head comparison before and after hair restoration. Left image shows significant hair loss; right image reveals restored hair density and full coverage. (Credited to Neil H. Riordan Center for Regenerative Medicine)

The surgical thickening was preceded by two S&V sessions, performed 3 months apart. The first hair transplant produced significant results (Fig. 7.15).

7.16 Scarring Alopecia: Case 9: S&V + FUE Treatments for Scarring from Skin Cancer Removal

Patient: Female, 63 years old.
Diagnosis: Removal of a basal cell carcinoma in the parietal area.
Treatments:

- 2021: S&V session
- 2022: FUE with 401 FUs.

Case Description:

This case involves scarring alopecia following surgical removal of a basal cell carcinoma in the parietal area.

The surgical thickening was preceded by an S&V session. The subsequent FUE hair transplant produced the results visible in the images below (Fig. 7.16).

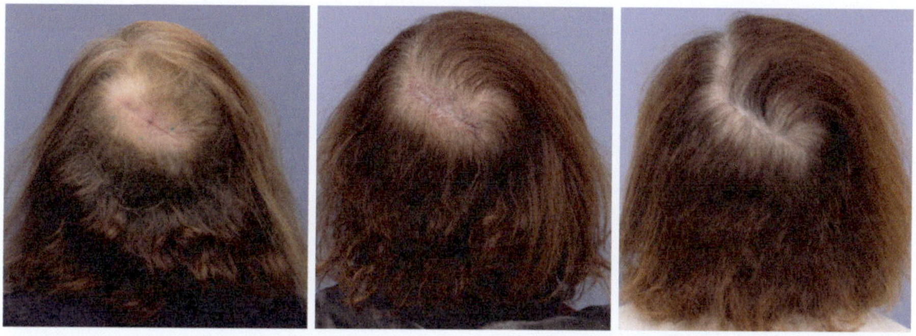

Fig. 7.16 Case 9 Sequential images showing progression of hair regrowth. From left to right: significant thinning, moderate improvement, and fuller hair coverage, demonstrating treatment effectiveness over time. (Credited to Peter Panagotacos)

7.17 Scarring Alopecia: Case 10: Trauma-Related Scarring Alopecia

Patient: Female, 47 years old.
Diagnosis: Scarring from trauma and subsequent healing by secondary intention.
Treatments:

- July 2020: first S&V session.
- October 2020: second S&V session.
- February 2021: FUE with 1700 FUs.

Case Description:

The patient experienced a severe road accident in 2019, resulting in a cranial trauma and the loss of a substantial portion of scalp tissue, approximately 20 cm^2, which was replaced using INTEGRA and healed by secondary intention.

Regenerative treatment with two S&V sessions followed by a FUE hair transplant enabled complete and satisfactory coverage of the scarred area, achieving a near-normal hair density (Fig. 7.17).

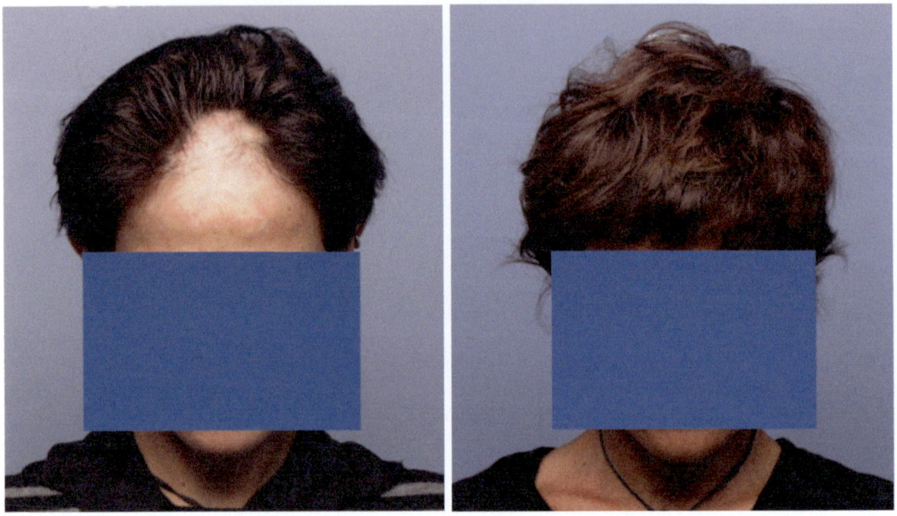

Fig. 7.17 Case 10 Before and after comparison of a hair transplant procedure. Left image shows thinning hair on the crown, while right image displays visibly fuller hair coverage posttreatment. (Credited to SMG Hair Transplant MD)

7.18 Scarring Alopecia: Case 11: Atrophic Scarring

Patient: Male, 16 years old.

Diagnosis: Necrosis from extravasation of embolization agent during treatment of an AVF (arteriovenous fistula).

Treatments:

- 2016: S&V session
- 2016: FUE with 875 FUs.

Case Description:

This case involves a scar located in the left frontal area, resulting from a previous embolization procedure for an arteriovenous fistula complicated by extravasation of the embolizing agent into the surrounding tissues.

The combined treatment with S&V and FUE hair transplantation achieved dense and natural coverage of the scar, with proper respect for the direction and angle of the hair. The scar was effectively concealed, as visible in the images below (Fig. 7.18).

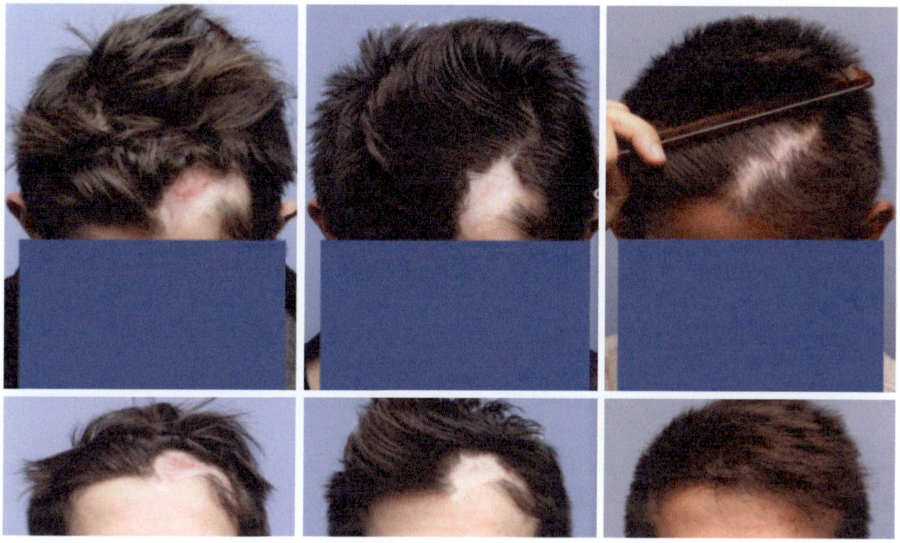

Fig. 7.18 Case 11 Four sequential top-view images showing stages of hair growth before and after hair restoration treatment, highlighting progressive improvement in scalp coverage. (Credited to Alvi Armani)

7.19 Conclusions

There is an ever-growing body of clinical evidence supporting the success of transplantation on scarred areas pretreated with AFT.

In nearly all cases treated, complete regrowth of the implanted follicular units has been observed. This success has gradually allowed the development of a new reconstructive procedure.

This procedure has over time supplanted surgical excisions, which often do not deliver the desired results, and skin expansion techniques, which are significantly influenced by the dynamics of flap rotation, where respecting the natural direction of hair growth proves impractical.

Today, the survival rates of follicular units after transplantation and their qualitative preservation are comparable to those achieved in procedures performed on healthy tissue. Results are considered highly satisfactory and proportional to the follicular units used. A high percentage of patients achieved definitive results in just one session.

From our clinical practice: although the exact regenerative mechanism through which the stem cell component of adipose tissue acts remains unclear, we have observed that treating the recipient tissue with AFT before transplantation has a remarkable effect on follicular structure.

These changes have had a significant impact on the quality and reliability of results and have undoubtedly contributed to the explosive demand for HT in cases of AGA. This increased demand outlines an “urgency” and the need for more active synergies among specialists in the reconstructive field.

The most evident difference compared to the numerous HT cases performed in the past years on scarred lesions has been the consistent and early regrowth of hair, which began between the second and third months after transplantation. It is unusual to observe the first results of the intervention as early as 4 months posttransplant (Figs. 7.12 and 7.13). This phenomenon is to be considered the result of a synergy of positive circumstances and actions and has often been the premise for excellent and stable results. The contrast with the past, characterized by interventions without preliminary regenerative operations, has represented, for the authors, a perfect litmus test on which it is possible to affirm that the benefits of these procedures are clear and evident.

Another advantage is the long-term durability of HT. This factor, often underestimated, is of fundamental importance. The immunomodulatory action provided by ADSCs and SVF fragments could allow greater predictability of long-term results, even for HT on primary scars (e.g., lichen planopilaris, AFF, etc.), which today are strongly influenced by the recurrence of the same pathology. The two cases included in this study were monitored for more than 3 years and showed a stable clinical picture.

Despite all the advantages of hair transplants on scarred lesions, it is essential that the surgeon fully understands the specific needs of women. A truly satisfactory result is not limited to restoring a minimal amount of hair but must instead aim to provide adequate volume and naturalness that restore femininity and confidence.

The analysis of the behavior of women who have experienced thinning induced by chemotherapy (CIA) highlights their distress when faced with significantly less dense hair compared to what they were accustomed to. In such situations, only a hair prosthesis, such as a wig, can restore an acceptable image.

These data provide important insights for deciding on the opportunity for an intervention, suggesting alternative techniques, particularly prosthetic ones. Our experience, therefore, leads us to evaluate surgical results in women with a history of scarring alopecia more carefully, as their expectations regarding hair density differ significantly from those of male patients.

Managing expectations must be a key aspect of any discussion with the patient, who must be well-informed about possible partial results before embarking on a surgical path. This conscious approach will allow the treatment to be undertaken with realistic expectations and greater serenity.

The results of the surgical plans executed were documented through photographs and dermatoscopy, with the addition, in many cases, of video recordings to offer a better understanding of the density obtained in dynamic situations.

Patient feedback has been overwhelmingly positive, even among those who had previously undergone complex surgical experiences due to underlying pathologies.

In conclusion, in the past, the surgeon did not have the appropriate tools to determine the correct number of grafts to be used in compromised areas. Unsatisfactory yields often led to frustrating results, aggravating the clinical situation due to the significant wastage of hair follicles, whose availability is limited. This dissatisfaction with partial results could have led patients to abandon their reconstructive path.

Today, thanks to scientific advances and modern techniques, it is possible to successfully address transplantation on scarred areas. The current approach allows these areas to be adequately measured and prepared, significantly reducing the margin of uncertainty and increasing the likelihood of achieving satisfactory and long-lasting final results. In this way, accurate planning and the precision of surgical techniques optimize the use of follicular units, ensuring a more natural aesthetic result and an overall more positive treatment experience for patients.

References

1. Fanti PA, Baraldi C, Misciali C, Piraccini BM. Cicatricial alopecia. G Ital Dermatol Venereol. 2018;153(2):230–42.
2. Somani N, Bergfeld WF. Cicatricial alopecia: classification and histopathology. Dermatol Ther. 2008;21(4):221–37.
3. Paganelli A, Contu L, Condorelli A, Ficarelli E, Motolese A, Paganelli R, Motolese A. Platelet-rich plasma (PRP) and adipose-derived stem cell (ADSC) therapy in the treatment of genital lichen Sclerosus: a comprehensive review. Int J Mol Sci. 2023;24(22):16107. https://doi.org/10.3390/ijms242216107.
4. Ding JY, Chen MJ, Wu LF, Shu GF, Fang SJ, Li ZY, Chu XR, Li XK, Wang ZG, Ji JS. Mesenchymal stem cell-derived extracellular vesicles in skin wound healing: roles, opportunities and challenges. Mil Med Res. 2023;10(1):36.
5. Piccinno MS, Petrachi T, Pignatti M, Murgia A, Grisendi G, Candini O, Resca E, et al. Human adipose mesenchymal stromal/stem cells improve fat transplantation performance. Cells. 2022;11(18):2799.
6. Gennai A, Bovani B, Colli M, Melfa F, et al. Evaluation of the number, biophysical and multipotent characteristics of adipose derived stem cells harvested by SEFFI procedure and interaction with different type of hyaluronic acids. Int J Regenerat Med. 2021;4:10.
7. Kumar AR, Ishii LE. Hair transplantation for scarring alopecia. Facial Plast Surg Clin North Am. 2020;28(2):177–9.
8. Yoo H, Moh J, Park JU. Treatment of postsurgical scalp scar deformity using follicular unit hair transplantation. Biomed Res Int. 2019;2019(13):3423657.
9. Loganathan E, Sarvajnamurthy S, Gorur D, Suresh DH, Siddaraju MN, Narasimhan RT. Complications of hair restoration surgery: a retrospective analysis. Int J Trichol. 2014;6(4):168–72.

MIX
Papier aus verantwortungsvollen Quellen
Paper from responsible sources
FSC® C105338

If you have any concerns about our products,
you can contact us on
ProductSafety@springernature.com

In case Publisher is established outside the EU,
the EU authorized representative is:
Springer Nature Customer Service Center GmbH
Europaplatz 3, 69115 Heidelberg, Germany

Printed by Libri Plureos GmbH
in Hamburg, Germany